[Coast Redwood]

COAST REDWOOD FOREST

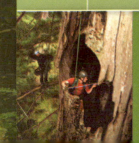

A coast redwood is an evergreen tree that can live up to 2,200 years. This type of tree is one of the tallest living things on Earth. It can grow up to 115.5 meters (379.1 feet) in height and 8 meters (26 feet) across. That can be as tall as a 30-story building. Its trunk is as wide as two cars.

Coast redwoods are native to California and Oregon, where they grow in large forests. Their bark is very thick, but soft, and their root system is shallow and wide spreading.

When forest fires burn through a group of coast redwoods, they can leave a large hollowed-out hole within the trunks. The hole provides rotten, spongy pockets in which plants and other living things can thrive.

NATIONAL GEOGRAPHIC
SCIENCE

LIFE SCIENCE

NATIONAL GEOGRAPHIC
School Publishing

PROGRAM AUTHORS

Randy Bell, Ph.D.

Malcolm B. Butler, Ph.D.

Kathy Cabe Trundle, Ph.D.

Judith S. Lederman, Ph.D.

David W. Moore, Ph.D.

Program Authors

RANDY BELL, PH.D.

Associate Professor of Science Education,
University of Virginia, Charlottesville, Virginia
SCIENCE

MALCOLM B. BUTLER, PH.D.

Associate Professor of Science Education,
University of South Florida, St. Petersburg, Florida
SCIENCE

KATHY CABE TRUNDLE, PH.D.

Associate Professor of Early Childhood Science
Education, The School of Teaching and Learning,
The Ohio State University, Columbus, Ohio
SCIENCE

JUDITH SWEENEY LEDERMAN, PH.D.

Director of Teacher Education,
Associate Professor of Science Education,
Department of Mathematics and Science Education,
Illinois Institute of Technology, Chicago, Illinois
SCIENCE

DAVID W. MOORE, PH.D.

Professor of Education,
College of Teacher Education and Leadership,
Arizona State University, Tempe, Arizona
LITERACY

Program Reviewers

Amani Abuhabsah
Teacher
Dawes Elementary
Chicago, IL

Maria Aida Alanis, Ph.D.
Elementary Science
Instructional Coordinator
Austin Independent
School District
Austin, TX

Jamillah Bakr
Science Mentor Teacher
Cambridge Public Schools
Cambridge, MA

Gwendolyn Battle-Lavert
Assistant Professor of Education
Indiana Wesleyan University
Marion, IN

Carmen Beadles
Retired Science Instructional
Coach
Dallas Independent School
District
Dallas, TX

Andrea Blake-Garrett, Ed.D.
Science Educational Consultant
Newark, NJ

Lori Bowen
Science Specialist
Fayette County Schools
Lexington, KY

Pamela Breitberg
Lead Science Teacher
Zapata Academy
Chicago, IL

Program Reviewers continued
on page iv.

Acknowledgments

Grateful acknowledgment is given to the authors, artists, photographers, museums, publishers, and agents for permission to reprint copyrighted material. Every effort has been made to secure the appropriate permission. If any omissions have been made or if corrections are required, please contact the Publisher.

Illustrator Credits
All illustrations by Precision Graphics.
All maps by Mapping Specialists.

Photographic Credits
Front Cover Andrew Geiger/Riser/
Getty Images.

Credits continue on page EM16.

Neither the Publisher nor the authors shall be liable for any damage that may be caused or sustained or result from conducting any of the activities in this publication without specifically following instructions, undertaking the activities without proper supervision, or failing to comply with the cautions contained herein.

The National Geographic Society
John M. Fahey, Jr.,
President & Chief Executive Officer

Gilbert M. Grosvenor,
Chairman of the Board

Copyright © 2011 The Hampton-Brown Company, Inc., a wholly owned subsidiary of the National Geographic Society, publishing under the imprints National Geographic School Publishing and Hampton-Brown.

National Geographic School Publishing
Hampton-Brown
www.myNGconnect.com

Printed in the USA.
RR Donnelley
Jefferson City, MO

ISBN: 978-0-7362-7710-5

11 12 13 14 15 16 17 18 19 20

2 3 4 5 6 7 8 9 10

Carol Brueggeman
K–5 Science/Math Resource
Teacher
District 11
Colorado Springs, CO

Miranda Carpenter
Teacher, MS Academy Leader
Imagine School
Bradenton, FL

Samuel Carpenter
Teacher
Coonley Elementary
Chicago, IL

Diane E. Comstock
Science Resource Teacher
Cheyenne Mountain School
District
Colorado Springs, CO

Kelly Culbert
K–5 Science Lab Teacher
Princeton Elementary
Orange County, FL

Karri Dawes
K–5 Science Instructional
Support Teacher
Garland Independent
School District
Garland, TX

Richard Day
Science Curriculum Specialist
Union Public Schools
Tulsa, OK

Michele DeMuro
Teacher/Educational Consultant
Monroe, NY

Richard Ellenburg
Science Lab Teacher
Camelot Elementary
Orlando, FL

Beth Faulkner
Brevard Public Schools
Elementary Training Cadre,
Science Point of Contact, Teacher,
NBCT
Apollo Elementary
Titusville, FL

Kim Feltre
Science Supervisor
Hillsborough School District
Newark, NJ

Judy Fisher
Elementary Curriculum
Coordinator
Virginia Beach Schools
Virginia Beach, VA

Anne Z. Fleming
Teacher
Coonley Elementary
Chicago, IL

Becky Gill, Ed.D.
Principal/Elementary Science
Coordinator
Hough Street Elementary
Barrington, IL

Rebecca Gorinac
Elementary Curriculum Director
Port Huron Area Schools
Port Huron, MI

Anne Grall Reichel Ed. D.
Educational Leadership/
Curriculum and Instruction
Consultant
Barrington, IL

Mary Haskins, Ph.D.
Professor of Biology
Rockhurst University
Kansas City, MO

Arlene Hayman
Teacher
Paradise Public School District
Las Vegas, NV

DeLene Hoffner
Science Specialist, Science
Methods Professor,
Regis University
Academy 20 School District
Colorado Springs, CO

Cindy Holman
District Science Resource
Teacher
Jefferson County Public Schools
Louisville, KY

Sarah E. Jesse
Instructional Specialist for
Hands-on Science
Rutherford County Schools
Murfreesboro, TN

Dianne Johnson
Science Curriculum Specialist
Buffalo City School District
Buffalo, NY

Kathleen Jordan
Teacher
Wolf Lake Elementary
Orlando, FL

Renee Kumiega
Teacher
Frontier Central School District
Hamburg, NY

Edel Maeder
K–12 Science Curriculum
Coordinator
Greece Central School District
North Greece, NY

Trish Meegan
Lead Teacher
Coonley Elementary
Chicago, IL

Donna Melpolder
Science Resource Teacher
Chatham County Schools
Chatham, NC

Melissa Mishovsky
Science Lab Teacher
Palmetto Elementary
Orlando, FL

Nancy Moore
Educational Consultant
Port Stanley, Ontario, Canada

Melissa Ray
Teacher
Tyler Run Elementary
Powell, OH

Shelley Reinacher
Science Coach
Auburndale Central Elementary
Auburndale, FL

Kevin J. Richard
Science Education Consultant,
Office of School Improvement
Michigan Department of
Education
Lansing, MI

Cathe Ritz
Teacher
Louis Agassiz Elementary
Cleveland, OH

Rose Sedely
Science Teacher
Eustis Heights Elementary
Eustis, FL

Robert Sotak, Ed.D.
Science Program Director,
Curriculum and Instruction
Everett Public Schools
Everett, WA

Karen Steele
Teacher
Salt Lake City School District
Salt Lake City, UT

Deborah S. Teuscher
Science Coach and
Planetarium Director
Metropolitan School District of
Pike Township
Indianapolis, IN

Michelle Thrift
Science Instructor
Durrance Elementary
Orlando, FL

Cathy Trent
Teacher
Ft. Myers Beach Elementary
Ft. Myers Beach, FL

Jennifer Turner
Teacher
PS 146
New York, NY

Flavia Valente
Teacher
Oak Hammock Elementary
Port St. Lucie, FL

Deborah Vannatter
District Coach, Science Specialist
Evansville Vanderburgh School
Corporation
Evansville, IN

Katherine White
Science Coordinator
Milton Hershey School
Hershey, PA

Sandy Yellenberg
Science Coordinator
Santa Clara County Office of
Education
Santa Clara, CA

Hillary Zeune de Soto
Science Strategist
Lunt Elementary
Las Vegas, NV

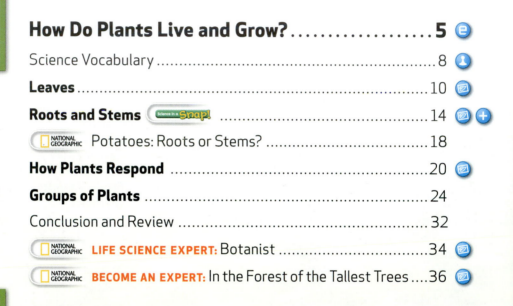

LIFE SCIENCE

CONTENTS

TECHTREK
myNGconnect.com

Student
eEdition

Vocabulary
Games

Digital
Library

Enrichment
Activities

TECHTREK
myNGconnect.com

Student
eEdition

Vocabulary
Games

Digital
Library

Enrichment
Activities

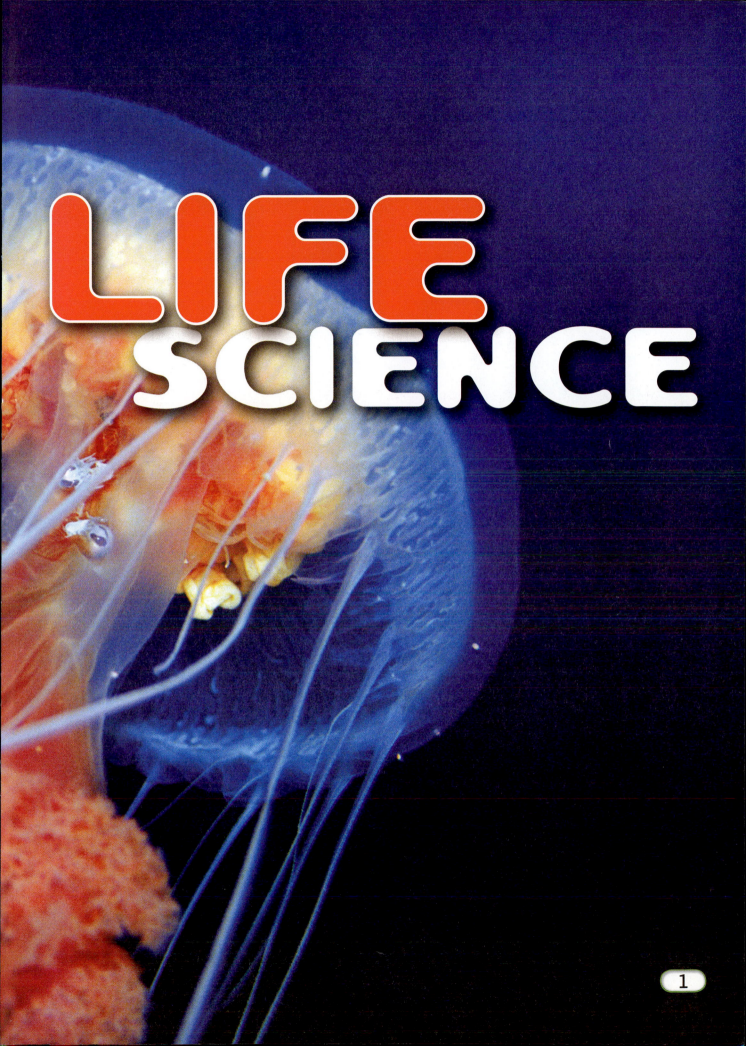

LIFE SCIENCE

What Is Life Science?

Life science is the study of all the living things around you and how they interact with one another and with the environment. This type of science investigates how living things are similar to and different from one another, how they live and reproduce, and how they function in the environment. Life science includes the study of humans, as well as all the other kinds of living things on Earth. People who study living things and the environment are called life scientists.

You will learn about these aspects of life science in this unit:

HOW DO PLANTS LIVE AND GROW?

Many different kinds of plants exist. Some have flowers, but not all of them do. A flower is just one kind of plant part that carries out a specific job as the plant grows. Life scientists study how plants live and grow.

HOW ARE ANIMALS ALIKE AND DIFFERENT?

Some kinds of animals have backbones while other kinds of animals do not. Animals can be grouped according to characteristics such as backbones. Life scientists study characteristics of animals and group them according to similarities and differences.

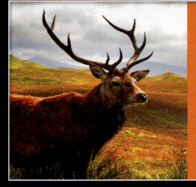

HOW DO PLANTS AND ANIMALS LIVE IN THEIR ENVIRONMENT?

Earth has both land and water environments. Plants and animals live together in their environment and depend on each other. Plants and animals are parts of food chains. Life scientists study how energy passes from plants to animals in a food chain.

HOW DO PLANTS AND ANIMALS SURVIVE?

Plants and animals have different features and behaviors that let them survive. Some plants and animals use camouflage. Other times body parts help animals. Body parts are adaptations. Plants have adaptations, too. Life scientists study how plants and animals use their parts to survive.

HOW DO PLANTS AND ANIMALS RESPOND TO SEASONS?

The seasons bring changes in light, temperature, and rainfall in most environments. Plants and animals respond to these seasonal changes in many different ways. Life scientists study how plants and animals respond to changes in the environment.

MEET A SCIENTIST

Tierney Thys: Marine Biologist, Filmmaker, Pilot

Tierney Thys is a marine biologist, filmmaker, pilot, and National Geographic Emerging Explorer. Since 2000, Tierney and her colleagues have been traveling the world ocean to study the giant ocean sunfish, or mola. Though these fish can grow more than three meters (ten feet) long and weigh over 2,270 kilograms (5,000 pounds), little is known about them. By placing high-tech satellite tags on molas and collecting mola tissue samples for genetic analysis, Tierney and her colleagues hope to uncover the mola's secrets.

"When it comes to fishes, the mola really pushes the boundary of fish form," says Tierney. "It seems a somewhat counterintuitive design for swimming in the waters of the open seas—a rather goofy design—and yet the more I learn about it, the more respect and admiration I have for it. That's what makes my work so exciting!"

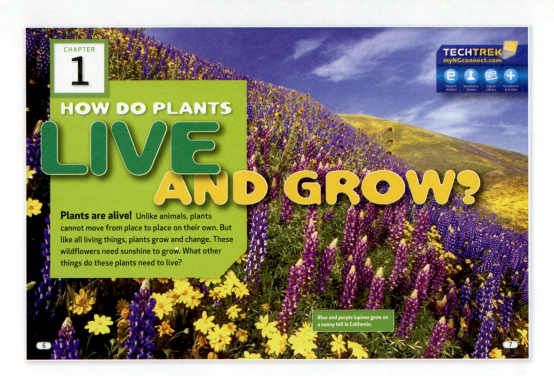

CHAPTER
1

HOW DO PLANTS LIVE AND GROW?

Plants are alive! Unlike animals, plants cannot move from place to place on their own. But like all living things, plants grow and change. These wildflowers need sunshine to grow. What other things do these plants need to live?

TECHTREK
myNGconnect.com

Blue and purple lupines grow on a sunny hill in California.

6

7

After reading Chapter 1, you will be able to:

- Identify the main parts of a plant, including leaves, roots, and stems. **LEAVES, ROOTS AND STEMS**

- Explain how leaves make food for the plant. **LEAVES**

- Explain how roots and stems move water and nutrients and support the plant. **ROOTS AND STEMS**

- Describe how plants respond to heat, light, and gravity. **HOW PLANTS RESPOND**

- Identify the stages in the life cycle of a flowering plant. **HOW PLANTS RESPOND, GROUPS OF PLANTS**

- Identify the characteristics used to classify the major groups of plants. **LEAVES, GROUPS OF PLANTS**

- Describe the reproductive structures of flowering plants, cone-bearing plants, ferns, and mosses. **GROUPS OF PLANTS**

- Explain that fossils give evidence about plants that lived long ago. **GROUPS OF PLANTS**

- **Science in a Snap!** Explain how roots and stems move water and nutrients and support the plant. **ROOTS AND STEMS**

CHAPTER

1

HOW DO PLANTS LIVE AND

Plants are alive! Unlike animals, plants cannot move from place to place on their own. But like all living things, plants grow and change. These wildflowers need sunshine to grow. What other things do these plants need to live?

GROW?

Blue and purple lupines grow on a sunny hill in California.

SCIENCE VOCABULARY

organism (OR-guh-niz-uhm)

An **organism** is a living thing. (p. 10)

A fern is an organism.

environment (en-VI-ruhn-ment)

The **environment** is all the living and nonliving things around an organism. (p. 20)

These pine trees live in a windy environment.

germinate (JUR-muh-NĀT)

Seeds **germinate** when they begin to grow. (p. 21)

A bean seed can germinate if the soil is moist and warm.

my
Science Vocabulary

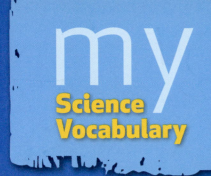

environment
(en-VI-ruhn-ment)

germinate
(JUR-muh-NĀT)

organism
(OR-guh-niz-uhm)

pollen
(POL-uhn)

reproduce
(rē-pru-DUS)

spore
(SPOR)

TECHTREK
myNGconnect.com

Vocabulary Games

---

reproduce (rē-pru-DUS)

To **reproduce** is to make more of its own kind. (p. 24)

> Apple trees reproduce by making seeds.

---

pollen (POL-uhn)

Pollen is a powder made by a male cone or the male parts of a flower. (p. 24)

> When bees visit flowers, pollen sticks to their bodies.

---

spore (SPOR)

A **spore** is a tiny part of a fern or moss that can grow into a new plant. (p. 28)

> The spores of a fern grow in cases often found underneath its leaves.

Leaves

Plants are living things, or **organisms** . Like all organisms, plants need water, food for energy, and space to live and grow. Plants also need air and nutrients from the soil. In addition, plants need the right temperatures.

Most plants have three main parts—leaves, roots, and stems. These parts work together to keep the plant alive.

Leaves use the energy of sunlight to make food for the plant.

How Plants Make Food Almost all plants make their own food. This is usually the job of the leaves. To make food, leaves need sunlight, air, and water.

Leaves capture the energy of sunlight. They use this energy to change air and water into food. Air enters the plant through tiny holes in its leaves. Water comes from the soil. Water travels to the leaves through the roots and stems.

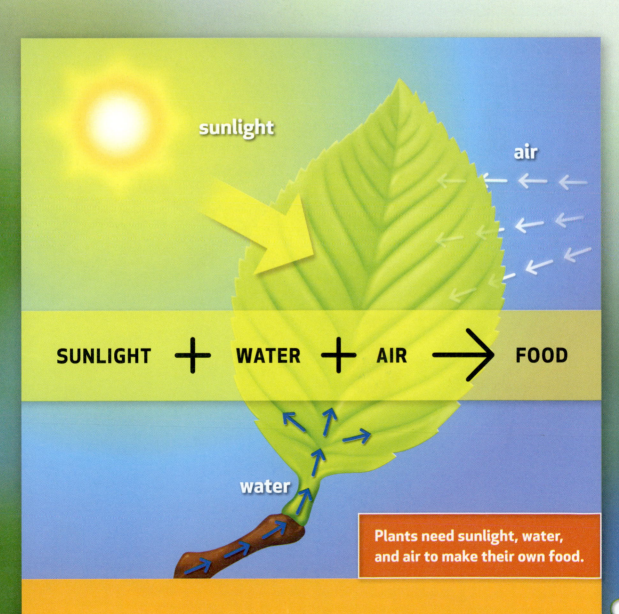

SUNLIGHT + WATER + AIR → FOOD

Plants need sunlight, water, and air to make their own food.

Kinds of Leaves
Each kind of plant has its own type of leaf. Compare the leaves in the picture below. Some of the leaves are big. Others are divided into many parts.

All leaves have veins that carry water through the leaf. Veins also carry food from the leaves to the rest of the plant. In some plants, the veins branch out in many directions. In other plants, the veins are straight and do not cross each other.

The leaves of flowering plants come in many shapes and sizes.

TYPES OF LEAVES

Most leaves are green. Different kinds of plants have leaves of very different shapes and sizes.

FLOWERING PLANTS
Many flowering plants have leaves that are broad and flat.

PLANTS WITH CONES
Pine trees and many other plants with cones have long, narrow leaves that are shaped like needles.

FERNS
The leaves of ferns are called fronds. Fronds are often divided into many parts.

MOSSES
The green parts of mosses are not true leaves because they do not have veins.

Before You Move On

1. What things do plants need to live and grow?
2. How do leaves make food for a plant?
3. **Analyze** Could the leaves of a plant live without roots and stems? Why?

Roots and Stems

Roots Which part of a plant is usually hidden from sight? The roots! Roots grow down into the soil. The main job of the roots is to take in water and nutrients. Nutrients are parts of soil that plants need to grow and stay healthy. Roots also hold the plant in place and help it stand upright.

The many small roots of grass help hold the soil in place.

Look at the roots in the picture below. Different plants have different kinds of roots. Grasses have many small roots. Other plants have one main root that grows deep into the soil. These long roots can reach water far below the surface. Water moves up the root to the plant's stems and leaves.

This sunflower has one main root that is long and thick.

main root

Stems Stems support the plant. They hold up the leaves and flowers. The trunk of a tree is a thick, woody stem. The picture shows the long trunks of palm trees. Notice how their trunks hold their leaves high above the ground. Other kinds of stems are slender, such as the stem of a daisy. The stalks of corn are also stems. Can you think of more kinds of stems?

TECHTREK
myNGconnect.com

Enrichment Activities

Stems carry water and food between the roots and leaves.

food

water

The trunks of these palm trees are stems.

Stems connect the roots and leaves. Stems carry water and nutrients from the roots to the leaves. Food made in the leaves moves through the stems to the rest of the plant.

Fill a cup about half full with water. Add about 20 drops of food coloring.

Your teacher will cut off the bottom of your flower stem. Place the stem in the cup. Wait one day. Then observe the flower.

What do you see? What happened to the colored water?

Before You Move On

1. What does a root take from the soil?
2. How do the roots of a plant get food? List the parts of a plant that food passes through as it travels to the roots.
3. **Generalize** How are the stalk of a bean plant and the trunk of a tree alike? How do a stalk and a trunk help a plant to live?

NATIONAL GEOGRAPHIC

POTATOES: ROOTS OR STEMS?

Do you like potatoes? Millions of people do!

Potatoes were first grown by people who lived in the Andes Mountains of South America. About 500 years ago, Spanish explorers brought potatoes to Europe. From Europe, farmers brought potatoes to many other parts of the world.

These farmers in Peru are harvesting potatoes.

Potatoes are still an important food for people in the Andes. They grow potatoes of many different shapes and sizes. Their potatoes also come in many different colors—green, yellow, red, and even purple!

Potatoes grow in the ground, but they are not roots. Potatoes are stems that store food for the plant. The leaves of a potato plant use sunlight to make food. Some of this food is then stored underground. When farmers harvest potatoes, you get to eat that food!

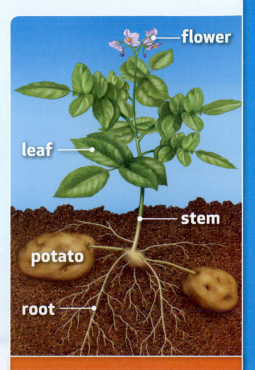

flower

leaf

stem

potato

root

Potatoes are stems that grow underground and store food for the plant.

How Plants Respond

Like all living things, plants respond to their environment. The **environment** of an organism is all the living and nonliving things around it. Temperature, light, and gravity are part of a plant's environment.

HOW A BEAN SEED GERMINATES

When the soil is warm enough, a bean seed will start to grow.

The bean seed takes in water from the soil and starts to swell.

The seed germinates, and the first root begins to grow downward.

The stem pushes upward. New roots grow.

Plants Respond to Heat Most plants grow only if the weather is warm enough. When the temperature rises in the spring, buds open and leaves begin to grow.

Many seeds respond to changes in the temperature of the soil. Some seeds germinate only when the soil is warm. When a seed **germinates** , it begins to grow. Because seeds do not germinate when it is cold, the young plants are not hurt by cold.

The seedling pushes out of the soil. The first leaves start to unfold.

The first leaves open.

Plants Respond to Light Look at the plant below. See how its stems are bending to the right. This plant is responding to the sunlight coming through the window. Plants get the energy they need by growing toward the light.

Some plants respond to changes in the direction of light during the day. Their leaves turn so they always face the sun. In the morning, they face east. In the evening, they face west.

The stems of this plant are growing toward the sunlight coming through the window.

Plants Respond to Gravity   Gravity is the force that pulls things toward Earth. Both roots and stems respond to gravity, but in opposite ways. Roots grow down and stems grow up.

Roots grow toward the pull of gravity. Growing down into the soil helps roots reach water and nutrients. Stems grow away from gravity. Growing upward helps stems reach sunlight.

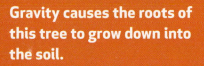

Gravity causes the roots of this tree to grow down into the soil.

Before You Move On

1. How do stems respond to light?
2. Contrast the way roots respond to gravity with the way stems respond to gravity.
3. **Infer**   In the early spring you plant some seeds. You water the seeds, but they do not grow. Why do you think the seeds did not germinate?

called **pollen**. When pollen lands on the female part of a flower, a fruit begins to grow. Inside the fruit are one or more seeds.

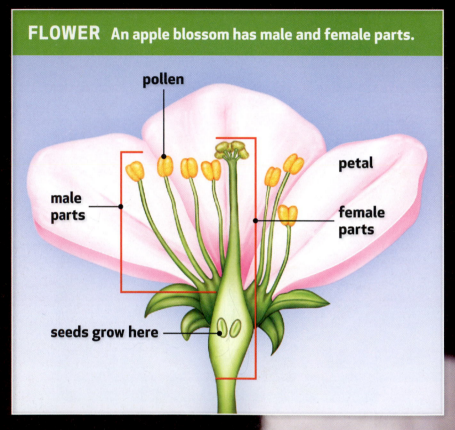

FLOWER An apple blossom has male and female parts.

pollen

petal

male
parts

female
parts

seeds grow here

POLLEN The male parts of a flower make pollen. Insects carry pollen to other flowers.

Each seed holds a young plant and a supply of food. The seed protects the young plant. The food helps the young plant start to grow.

Some flowering plants, such as apple trees, live for many years. Each year they make more seeds. Other flowering plants live for only one year. After they make seeds, they die.

LIFE CYCLE OF AN **APPLE TREE**

seed

SEEDLING When a seed falls on the ground, it germinates and grows into a young plant.

FRUIT When pollen reaches the female part of a flower, an apple begins to grow. Inside the apple are several seeds.

FLOWERING TREE After many years, the seedling grows into a tree. In the spring, the tree flowers.

Plants with Cones   How is a pine tree different from an apple tree? One important difference is that a pine tree has cones and an apple tree has apples! The seeds of a pine tree grow inside cones. Plants with cones do not have flowers or fruits.

Pine trees have two kinds of cones. Male cones make pollen. Wind blows the pollen to female cones. Then seeds grow inside the female cones. When the seeds are ripe, the scales of the female cones open and the seeds blow away.

These pitch pines are growing in Acadia National Park in Maine.

Like plants with flowers, plants with cones have roots, stems, and leaves. Many plants with cones have long thin leaves called needles. Pine needles stay on the tree all year, even in winter. Needles are covered by a thin layer of wax that protects them in cold weather.

Pitch pines have male and female cones. Male pine cones make pollen.

Seeds grow inside the female cones. When the cones open, the seeds blow away.

Plants Without Seeds Mosses and ferns do not have seeds. Instead, they reproduce with spores. A spore is a tiny part of a moss or fern that can grow into a new plant. Spores do not have a supply of food for the young plant.

Ferns usually have big leaves divided into many smaller parts. On the underside of the leaves are cases where spores grow.

The big, lacy leaves of ferns are called fronds.

Underneath the leaves of a fern are cases where spores grow.

Each case holds many tiny spores.

spores

Like ferns, mosses grow where it is shady and damp. Mosses do not have roots and stems, so they are usually much smaller than ferns. The spores of mosses form on tiny stalks that grow out of the green part of the plant.

The spores of mosses grow inside capsules at the top of slender stalks.

Grouping Plants That Lived Long Ago Fossils are the remains of organisms that lived a very long time ago. The picture shows fossils of tree trunks from an ancient forest. Other fossils have formed from the remains of leaves, stems, and even flowers!

 These fossil tree trunks are found in the Petrified Forest National Park in Arizona. The trees that left these fossils lived millions of years ago.

Fossils show that many ancient plants were similar to plants that are alive today. Some ancient plants had flowers and fruits. Other ancient plants had seeds in cones. And some ancient plants did not make seeds at all. Scientists group some fossils together with plants that are alive today.

Look at the two fossils shown here. Which fossil came from a flowering plant? Which one came from a fern?

Before You Move On

1. What is pollen?
2. List these steps in the life cycle of a flowering plant in the correct order: death, growth, germination, reproduction. Start with germination.
3. **Infer** Fir trees have leaves shaped like thin needles. Do you think fir trees have flowers? Explain why.

Conclusion

Plants need food, water, nutrients, and space to live and grow. Leaves use water, air, and sunlight to make food. Roots take in water and nutrients. Stems support the plant and carry water and food between the roots and leaves. Plants respond to heat, light, and gravity. Many plants reproduce by making seeds in flowers or cones. Other plants reproduce by making spores.

Big Idea Plants have different parts that work together to help them live, grow, and reproduce.

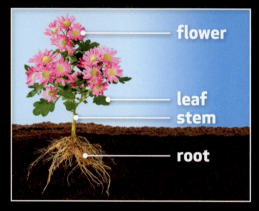

Vocabulary Review

Match the following terms with the correct definition.

A. organism	**1.** To begin to grow
B. environment	**2.** To make more of its own kind
C. germinate	**3.** A living thing
D. reproduce	**4.** A powder that is made by the male part of a flower or male cone
E. pollen	**5.** A tiny part of a moss or a fern that can grow into a new plant
F. spore	**6.** All the living and nonliving things around an organism

Big Idea Review

1. **Describe** Describe two ways that roots help a plant live and grow.

2. **Identify** The four main groups of plants are ferns, mosses, plants with cones, and plants with flowers. Which of these groups of plants reproduce with seeds? Which reproduce with spores?

3. **Summarize** How do the roots, stems, and leaves work together to make food for the plant? Begin by explaining what things a leaf needs to make food.

4. **Classify** A scientist discovers the fossil of a plant that has a fruit. In what group of plants would the scientist classify this plant? Explain why.

5. **Predict** When the weather becomes warm in the spring, how will the buds on a tree respond?

6. **Analyze** How do stems and leaves respond to light? How does this response help a plant to grow?

Write About Plants

Infer Look at the picture. Why do you think the stem of the plant is bent? Write what you think happened to the plant. Also describe how the roots are growing. Explain why the stem and roots are growing this way.

CHAPTER **1**

LIFE SCIENCE EXPERT: BOTANIST

Grace Gobbo, Botanist

Do you enjoy learning about plants? Grace Gobbo does. She is a botanist, a scientist who studies plants. Her goal is to help save plants that are important to her community.

Grace lives in Tanzania in East Africa. Tanzania is a beautiful country with forests, grasslands, and mountains. It has more than 10,000 kinds of plants! Grace visits different places in Tanzania to learn about the plants that grow there.

TECHTREK
myNGconnect.com

Digital Library

Grace Gobbo studies trees, flowers, fruits, and vines. She is looking for plants that can cure diseases.

TECHTREK
myNGconnect.com

Student
eEdition

Digital
Library

Grace studies how people in her country use plants. She visits many different villages. There she asks the elders which plants they use to treat diseases. Then she writes down what the elders say. In the past, nothing was written down. Grace is helping to preserve their knowledge. Grace also works with people in the villages to help preserve the forests and other wild places.

These plants grow in tropical forests on Mount Kilimanjaro in Tanzania.

Tanzania is a country in East Africa. Thousands of different plants grow there.

BECOME AN EXPERT

In the Forest of the Tallest Trees

Have you ever seen a forest with trees that are taller than 30-story buildings? You would if you visited the redwood forests of California and Oregon.

The Pacific Coast has just the right **environment** for redwood trees to grow. The ocean brings cool, rainy weather. Fog often fills the air.

Redwood trees grow in parts of California and Oregon.

environment

An **environment** is all the living and nonliving things around an organism.

TECHTREK
myNGconnect.com

e
Student
eEdition

Digital
Library

Redwoods are the tallest trees in the world. The trunk of a redwood has no elevators, but materials move up and down inside it. Its roots take in water and nutrients, which travel up the trunk to the leaves. The leaves use water, air, and sunlight to make food for the tree. Food travels from the leaves all the way down to the roots!

Old tree
1,500 years old

Trees less than
100 years old

Humans

Even the tallest human is tiny compared to a redwood tree.

Old Redwoods, New Redwoods

Many redwood trees are 500 years old. Some have lived for 2,000 years! Yet every redwood began life as a seed.

Redwood seeds grow inside cones. Redwood trees have two kinds of cones: male and female. Male cones make a powder called <mark>pollen</mark>. If pollen lands on a female cone, seeds will form.

Redwood trees have male and female cones. Seeds grow in the female cones.

female cones

male cones

TECHTREK
myNGconnect.com

Digital Library

If a seed lands in a good place, it will germinate.

pollen

Pollen is a powder made by a male cone or the male parts of a flower.

When the cone opens, its seeds fall to the ground. If a seed lands in a good place, it will start to grow, or **germinate** .

After many, many years, the seedling will grow into a tall redwood. The new tree will be similar to its parents. But each tree will also have its own special shape and pattern of branches.

After many years, a redwood tree will begin making cones.

germinate

When seeds **germinate,** they begin to grow into new plants.

Other Plants of the Forest

Many other kinds of plants also grow in the redwood forest. Most of these plants reproduce with flowers. The flowers of dogwoods and rhododendrons are very beautiful.

Rhododendrons flower in the spring.

reproduce

To **reproduce** is to make more of its own kind.

The seeds of flowering plants grow in fruits. The acorns of tanoaks are fruits. The sweet berries of huckleberry bushes are also fruits.

The shade of the tall trees makes the forest dark. But the damp forest floor is a good place for mosses and ferns to grow. Mosses and ferns reproduce by making **spores** .

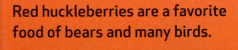

Red huckleberries are a favorite food of bears and many birds.

Mosses and ferns often grow on fallen logs.

spore

A **spore** is a tiny part of a fern or moss that can grow into a new plant.

A Community High in the Trees

If you climbed up a redwood tree, what would you find? An amazing number of organisms. An **organism** is a living thing.

Stellar's jay

This trunk of a redwood is growing up from the branch.

leatherleaf fern

The marbled murrelet is a bird that feeds on fish in the ocean, but nests high up in old trees.

Salamanders look for food among the dead leaves.

organism

An **organism** is a living thing.

Soil has formed on some of the old branches. The art shows some of the plants and animals that live on these branches.

red huckleberry

yellow-cheeked chipmunk

A thick mat of soil has formed on this branch. The roots of many plants grow through it. Tiny animals called copepods live in the moist soil.

evergreen huckleberry

CHAPTER 1

SHARE AND COMPARE

Turn and Talk How do the roots, stems, and leaves of a redwood tree work together to help it live and grow? Work with a partner to form a complete answer to this question.

Read Select two pages from this section. Practice reading the pages. Then read them aloud to a partner. Talk about why the pages are interesting.

my SCIENCE notebook **Write** Write a conclusion about how the plants in the redwood forest live and grow. State what you think is the Big Idea of this section. Share what you wrote with a classmate. Compare your conclusions.

my SCIENCE notebook **Draw** Draw a picture of one plant or animal that lives in the redwood forest. Add labels or write a caption to explain what you drew. Combine your drawing with those of your classmates to make a mural of the redwood forest.

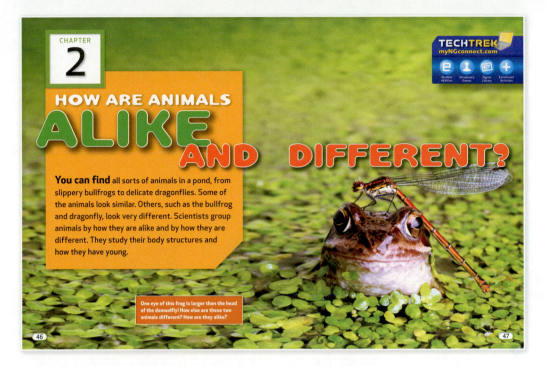

46

CHAPTER 2

HOW ARE ANIMALS ALIKE AND DIFFERENT?

You can find all sorts of animals in a pond, from slippery bullfrogs to delicate dragonflies. Some of the animals look similar. Others, such as the bullfrog and dragonfly, look very different. Scientists group animals by how they are alike and by how they are different. They study their body structures and how they have young.

One eye of this frog is larger than the head of the damselfly! How else are these two animals different? How are they alike?

47

After reading Chapter 2, you will be able to:

- Classify animals based on observable physical characteristics. **GROUPING ANIMALS; ANIMALS WITHOUT BACKBONES; FISHES, AMPHIBIANS, AND REPTILES; BIRDS AND MAMMALS**

- Compare life cycles of different animals and how animals closely resemble their parents and others in their species. **ANIMAL LIFE CYCLES**

- Observe and explore how fossils provide evidence about animals that lived long ago and the nature of the environment at that time. **CLUES FROM FOSSILS**

- **Science in a Snap!** Classify animals based on observable physical characteristics. **BIRDS AND MAMMALS**

HOW ARE ANIMALS ALIKE AND

You can find all sorts of animals in a pond, from slippery bullfrogs to delicate dragonflies. Some of the animals look similar. Others, such as the bullfrog and dragonfly, look very different. Scientists group animals by how they are alike and by how they are different. They study their body structures and how they have young.

One eye of this frog is larger than the head of the damselfly! How else are these two animals different? How are they alike?

DIFFERENT?

SCIENCE VOCABULARY

classify (CLA-si-fī)

To **classify** is to place into groups based on characteristics. (p. 50)

Scientists classify crabs in the same group as spiders.

backbone (BAK-bōn)

A **backbone** is a string of separate bones that fit together to protect the main nerve cord in some animals. (p. 51)

A cheetah's backbone bends when the cheetah runs.

backbone

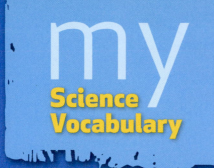

my Science Vocabulary

backbone (BAK-bōn)

classify (CLA-si-fī)

invertebrate (in-VUR-tuh-brit)

vertebrate (VUR-tuh-brit)

TECHTREK
myNGconnect.com

Vocabulary Games

vertebrate (VUR-tuh-brit)

A **vertebrate** is an animal with a backbone. (p. 52)

A backbone helps many vertebrates walk, run, fly, jump, or swim.

invertebrate (in-VUR-tuh-brit)

An **invertebrate** is an animal without a backbone. (p. 53)

A spiny lobster is an invertebrate.

49

Grouping Animals

Animals come in many shapes, sizes, and colors. How do scientists sort them? One way scientists **classify**, or group, animals is by their characteristics, or features and behaviors. These crabs all have ten legs and hard outer coverings. The iguanas have four legs and scaly skin. Classifying helps scientists better understand animals.

These iguanas and crabs live on the Galápagos Islands off the coast of South America. What characteristics do you observe about them?

One characteristic that scientists use to classify animals is whether or not an animal has a backbone. A **backbone** is a string of separate bones that fit together to protect the main nerve cord in some animals. The head and other bones are connected to the backbone. It helps the animal move.

These marine iguanas have a backbone. The Sally lightfoot crabs do not.

Animals with backbones are called vertebrates . The cheetah is just one kind of vertebrate. Catfish, tree frogs, king snakes, gulls, and bats are vertebrates, too. Even though each kind of animal is very different from the other, they all have backbones in about the same location as the cheetah's. The backbone is part of a vertebrate's skeleton.

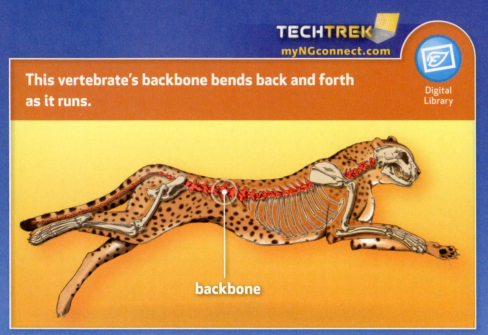

This vertebrate's backbone bends back and forth as it runs.

backbone

Animals without backbones are called invertebrates . Some invertebrates, such as earthworms, have no skeleton at all. Others such as clams grow a protective shell on the outside of their bodies. Still others such as lobsters and butterflies grow a hard outer covering that is like an outside skeleton. But none has a backbone.

This invertebrate's skeleton is on the outside of its body. It does not have a backbone.

Before You Move On

1. Why do scientists classify animals?
2. Observe the invertebrates and the vertebrates. How are they alike? How are they different?
3. **Apply** What are some kinds of animals that you think have a backbone?

Animals Without Backbones

You might think that vertebrates are the most common kinds of animals on Earth because you notice them more often. But most kinds of animals are invertebrates. Some live on land, while others live in water. Many are smaller than your fingertip, but some, such as the giant squid, are among the largest animals on Earth.

JELLIES
These animals have round bodies with tentacles they use to sting the living things they eat. Jellies move from place to place.

Look at the animals in the photos on these pages. Each animal belongs to a different invertebrate group. How do the animals differ from one another? How are they alike?

SPONGES
Sponges live on the ocean floor. They do not move around. They get food from the water that passes through their many pores, or holes.

WORMS
Some worms are flat, some are round, and others have body sections. They can live on land, in ponds, in other animals, or, like this flatworm, on the ocean floor.

MOLLUSKS
Mollusks have a soft body. Most have a shell that can be inside or outside, such as this snail's shell. They also have a muscular foot for moving.

ARTHROPODS
The bodies of all arthropods have a hard outer covering. Arthropods also have jointed legs and a body divided into sections.

SEA STARS
Sea stars and other animals like them have spiny skin. Many also have suction-cup-like feet that help them eat and move.

Of all the different kinds of invertebrates, most are insects. Scientists classify arthropods with three body sections and six legs attached to the middle section as insects. Most insects also have wings, antennae for sensing the environment, and eyes made up of several smaller ones.

Find the spicebush swallowtail's antennae.

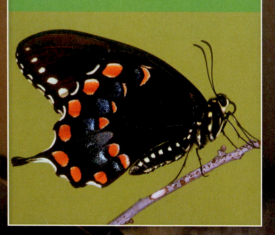

Notice that you can see through the wings on the bar winged skimmer.

Each big eye of this yellowjacket wasp is made of several hundred small ones. It can see close-by movements very well.

You might call all the insects on these pages "bugs." People commonly do. Some insects even have "bug" in their names. The milkweed bug in the photo below is an insect. Scientists can use the milkweed bug's wings and mouthparts to classify it as a certain kind of insect.

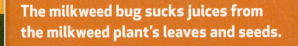

The milkweed bug sucks juices from the milkweed plant's leaves and seeds.

Before You Move On

1. What is the largest group of invertebrates?
2. Is a jellyfish a vertebrate or invertebrate? Explain how you know.
3. **Analyze** How does the body structure of different arthropods help you classify them?

Fishes, Amphibians, and Reptiles

Fishes Scientists classify animals as fish using four characteristics. Fishes are vertebrates that live in water and have fins, scales, and gills. Fishes use gills to take in oxygen from the water. Fishes take in water through their mouths. Then the water passes out over the gills. As the water passes over the gills, oxygen moves from the water into the fish's blood.

gills

fin

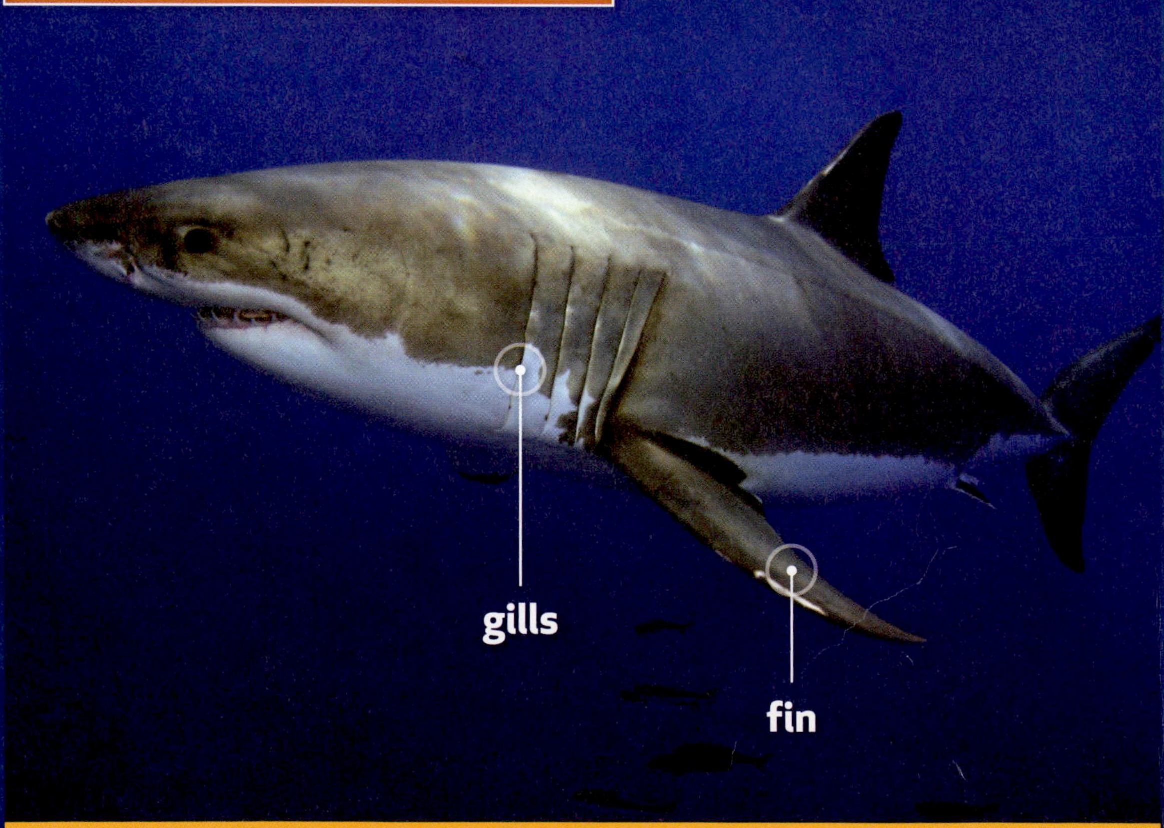

Fishes can look very different from one another. A shark is a fish. Sharks have scales that are rough as sandpaper. You can see the slits where the water moves out over the gills. Most fishes are like the koi. Its scales feel slimy. It also has a covering over the gills so you can't see the gills on the outside of the fish's body.

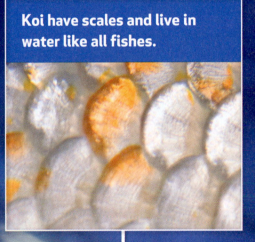

Koi have scales and live in water like all fishes.

Amphibians Have you ever seen a frog or a toad? Frogs and toads are part of another group of vertebrates called amphibians. Newts and salamanders are also part of this group. A young amphibian lives in water and breathes with gills like a fish. Most amphibians then grow lungs and legs, and they can live on land.

This Eastern, or red spotted, newt is almost an adult. It has grown legs and lungs, and lives on land.

The barred tiger salamander grows very large for a salamander. It can be as big as your outstretched hand!

Toads and frogs look very similar. Why do scientists classify toads and frogs as different kinds of amphibians? Toads have shorter back legs. They walk or hop instead of jumping and swimming. Toad skin also feels drier. But they keep it wet by rubbing water from shallow pools or dew over their backs. And, like frogs, toads have teeth. But only on the bottom jaw.

African bullfrogs are among the largest frogs. They can weigh almost two kilograms (about five pounds).

The warty skin of the American toad produces a milky poison that makes predators sick. The male puffs out its throat when it calls as it tries to attract a mate.

Reptiles You may have seen members of another group of vertebrates called reptiles. Snakes, turtles, and lizards are all reptiles. Alligators and crocodiles are reptiles, too.

Unlike fish and amphibians, reptiles mostly lay their eggs on land. Reptile eggs have a soft but tough shell that protects them. When the eggs hatch, the young reptiles look like their parents. They have lungs and breathe air.

Compare this collared lizard to the reptiles on the next page. How are they alike? How are they different?

Reptiles have a covering of hard scales that is not slimy. Scales help keep their bodies from drying out. The scales also protect reptiles from other animals that may try to eat them. Scales can also help reptiles move.

Like other snakes, the emerald tree boa has special scales on its belly that help it climb trees.

Like other turtles, the backbone of this box turtle is part of the shell, which is covered with special scales.

Before You Move On

1. What are the characteristics of a fish?
2. What characteristic do fish and young amphibians have in common?
3. **Generalize** How are a newt and a lizard alike? How are they different?

Birds and Mammals

Birds What do you notice about the vertebrates on this page? While they look very different, they both share certain characteristics. Scientists classify vertebrates with feathers, wings, and two legs covered with scales as birds.

The beak and legs of the scarlet ibis are longer than many birds. It wades in water to catch food.

No other kind of animal has feathers. These help birds fly and stay warm. Birds puff out their feathers to trap air when they are cold. The trapped air next to their skin warms and helps the birds stay warm. But not all birds fly. The feathers of a flying bird such as the ibis have a different structure than the feathers of a penguin, which doesn't fly.

Even though all birds have wings, they do not all fly. The rockhopper penguin hops to where it needs to be!

Mammals Vertebrates that have hair or fur and make milk to feed their young are called mammals. You cannot always see the hair or fur on a mammal. Some mammals, such as whales, lose their hair soon after they are born.

Most mammals take care of their young after they are born. Female mammals make milk for their young until the young are old enough to eat other foods.

Elephants are mammals so they have hair. But the hair is not as important for keeping warm as in other mammals.

Mammals, too, are very different from one another. Some have long trunks, some live in the ocean, and some can fly! Scientists use many characteristics and behaviors to classify mammals.

TECHTREK
myNGconnect.com

Science in a Snap! Classify Animals

Enrichment Activities

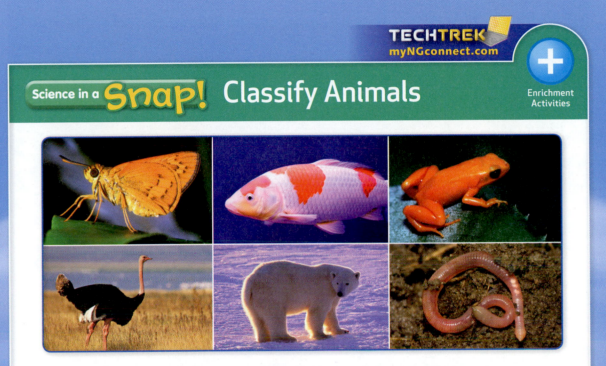

Observe the pictures above. Classify each animal as a vertebrate or an invertebrate. Organize your classifications in a chart.

Share your work with your partner.

Before You Move On

1. What characteristic do birds have that no other vertebrate has?
2. What two characteristics make mammals different from other vertebrates?
3. **Apply** Name five other kinds of animals that you think are mammals.

Animal Life Cycles

Scientists observe animals' life cycles, or how they grow and reproduce. Most kinds of animals have young by laying eggs. In some kinds of animals, the adult dies very soon after the eggs are laid. In other kinds of animals, the adult lives a long time and can reproduce many times before it dies.

An adult pipevine swallowtail butterfly lives several days. It mates, lays eggs, and then dies.

Life cycles have different stages or parts. Many insects have a four-stage life cycle in which the animal looks very different at each stage. Look at the life cycle of the pipevine swallowtail butterfly. The adult lays eggs about ten days after it breaks out of the pupa. It dies after it lays the eggs. But the eggs hatch and start the life cycle over again.

LIFE CYCLE OF A **BUTTERFLY**

LARVA This stage is often called a caterpillar.

PUPA A butterfly pupa is called a chrysalis.

EGGS The adult lays the eggs on the underneath side of pipevine leaves.

ADULT Adults feed on nectar from plants, such as thistles, lilacs, and phlox.

Frogs and other amphibians go through a life cycle where the animal looks different at different stages. Trace the life cycle diagram with your finger. Adult amphibians can live for many years. They can lay eggs at least once each year at a certain time for that kind of amphibian.

LIFE CYCLE OF A **FROG**

EARLY TADPOLE
In this stage, the young leopard frog has no legs and breathes through gills like a fish.

LATE TADPOLE
After the front legs form, the change to adult is about complete.

EGGS Adult leopard frogs lay up to 3000 eggs at once!

ADULT Leopard frogs do not lay eggs until they are 2 to 3 years old.

Most amphibians hatch from eggs in the water. Young amphibians live in water and breathe with gills. Their bodies change as they grow into adults. They grow legs, and their tails may shrink. Most amphibians also grow lungs. With these changes, amphibians can live on land.

Leopard frogs use their strong hind legs to escape danger. They live about seven years.

The life cycle of mammals is very different from that of insects and amphibians. Most mammals do not lay eggs. The young develop inside the mother and are born alive. Mammals can live for many years and reproduce many times. The young cannot take care of themselves. They drink milk from their mother until they can eat other foods.

For the next three weeks the mother Thomson's gazelle will hide her fawn in tall grass. She will return twice a day to feed the fawn milk.

The young grow and change, becoming larger and stronger. Many times the young learn behaviors from the parents, such as how to find food and how to survive on their own. Over time, the young become adults. Then they may reproduce to make more of their own kind.

These young American bison will drink milk from their mothers for about 1 year. They can have young at about age 3 and will live for about 20 years.

Before You Move On

1. What is a life cycle?
2. Where can baby animals develop?
3. **Draw Conclusions** Dolphins are a large animal that lives in the ocean. They have young that are born alive. Do you think dolphins are most likely a fish or a mammal? Explain.

Clues from Fossils

Many kinds of animals no longer live on Earth, but scientists can still classify them by observing their fossils. Fossils are the traces of living things that lived long ago. A fossil can be part of a living thing, such as a bone or a tooth. It can also be a mark, such as a footprint, that formed as mud turned into rock over millions of years.

gill cover

backbone

fin

This fossil formed millions of years ago. Look at the parts that are preserved. Infer why you see these parts preserved and not others, such as scales.

The characteristics in fossils can show how animals from long ago are similar to those that live today. The backbone of the fossil in the big photo shows that this animal was a vertebrate. What other traits shown in the fossil give hints about how the animal could be classified?

Compare the fossil to this photo of the skeleton of a fish that lives today.

Fossils tell scientists what the environment was like long ago. Look at the big photo. What does the environment look like? In this area of Indiana, scientists have found fossils of huge trees that could grow only in very warm, swampy environments. They have also found fossils of animals that could live only in warm, shallow oceans.

This crinoid fossil was found near Crawfordsville, Indiana. Most scientists group this animal with sea stars.

What do these fossils tell scientists about this area? They are clues that the environment here has changed over time. Now the area is forests and farmland with warm summers and cold winters. Millions of years ago it was covered with shallow ocean water and the weather was very warm all year long.

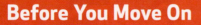

This fossil tree bark is from a tree that lived at the same time as crinoids. It grew as tall as a three-story building. It has many characteristics of a modern plant that grows only about as tall as your knee.

Before You Move On

1. What is a fossil?
2. What can fossils show about an animal's environment?
3. **Analyze** Some scientists think crinoids might be related to jellies. What body part might support this idea?

NATIONAL GEOGRAPHIC

LIFE ON THE REEF

A coral reef is like a small underwater island. Invertebrates called coral form a reef as they grow on the ocean floor. The Great Barrier Reef, located off the coast of Australia, is the world's largest coral reef. Scientists explore the reef to observe and study the animals. Sometimes they even find kinds they haven't seen before!

Scientists use the characteristics of living things to study and classify animals.

Many different kinds of animals live on a coral reef. Scientists use cameras and other tools to observe the animals. Animals can be hard to find because they hide in the coral and live in holes left when coral dies. Scientists try to classify any new kinds of animals they haven't seen before by comparing their characteristics with groups of living things they already know.

This reptile is a sea snake. What are some of its characteristics?

What arthropod characteristics do you see in this green-banded snapping shrimp?

What animals might scientists compare this sea horse with?

Conclusion

Classifying animals helps scientists to study and better understand them. Scientists observe animals' characteristics. Invertebrates are animals that do not have backbones and vertebrates are animals with backbones. Insects are the largest group of invertebrates. Vertebrates include fishes, amphibians, reptiles, birds, and mammals. Some animals reproduce by laying eggs while others have young that are born alive. Fossils tell about plants, animals, and the environment of long ago.

Big Idea Animals can be classified into groups based on their characteristics and behaviors.

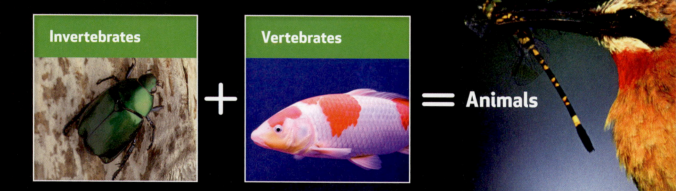

Invertebrates + Vertebrates = Animals

Vocabulary Review

Match the following terms with the correct definition.

A. classify

B. vertebrate

C. invertebrate

D. backbone

1. An animal without a backbone

2. A string of separate bones that fit together to protect the main nerve cord in some animals

3. An animal with a backbone

4. To place into groups based on characteristics

Big Idea Review

1. Name What are some characteristics of insects?

2. Define What is a vertebrate?

3. Explain How can a fossil animal tell about what the environment was like in the past?

4. Compare and Contrast Compare how young develop in insects, amphibians, and mammals.

5. Draw Conclusions Bats are animals that fly. They are covered in hair and make milk for young. Do you think they are birds or mammals? Give reasons.

6. Generalize The word *amphibian* comes from old words that mean "double" and "life." Why is this a good name for frogs, toads, salamanders, and newts?

Write About Animals

Classify This young deer probably still drinks milk from its mother. Write a few sentences that tell what characteristics you could use to classify the deer and the butterfly. Then tell what groups you would classify them in.

CHAPTER 2 · LIFE SCIENCE EXPERT: VETERINARIAN

Carlos Sanchez, Zoo Veterinarian

A veterinarian—or vet, for short— is a doctor for animals. A vet helps animals stay healthy and treats them when they are sick. In school, vets study all about animals. They learn about the diseases that animals get and how to treat those diseases. Many vets do surgery too.

Pandas live in only a few zoos in the United States. Dr. Sanchez is one of the few vets who treats them.

Dr. Sanchez views several radiographs of a species of monkey called the Golden lion tamarin.

TECHTREK
myNGconnect.com

e
Student
eEdition

Digital
Library

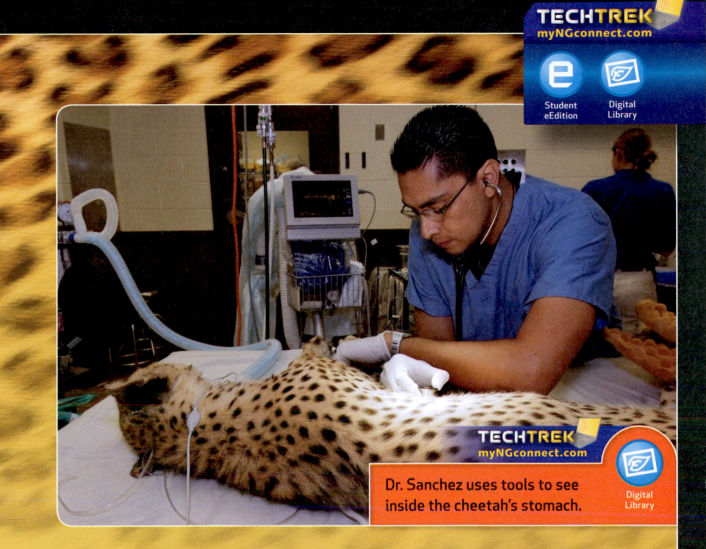

Dr. Sanchez uses tools to see inside the cheetah's stomach.

Many vets treat cats, dogs, and other pets. Some vets care for farm animals, such as horses and cows. But Dr. Carlos Sanchez treats more unusual animals. He is a vet at the Smithsonian's National Zoo in Washington, D.C. He treats all of the zoo's animals.

Dr. Sanchez does not work only at the zoo. He has traveled to China to study pandas. He has traveled to Africa to study giraffes and cheetahs. Dr. Sanchez shares what he learns so that everyone benefits from his trips.

When Dr. Sanchez was in grade school, he knew he wanted to be a vet someday. His advice for young people is to study hard. Maybe they will operate on an elephant some day!

BECOME AN EXPERT

The Florida Everglades:
More Than Meets the Eye

The Everglades is a national park that covers the southern tip of Florida. The land here is flat, wet, and has many different habitats, or areas where animals can live. This environment supports many kinds of animals.

Everglades National Park is often called the river of grass.

Everglades National Park

Invertebrates called arthropods live in the Everglades. Insects are one type of arthropod. Like all arthropods, insects have a body divided in sections, jointed legs, and a hard outer skeleton. Insects have 3 body sections and 6 legs. They can be further classified into groups based on other traits such as mouth parts.

SOME EVERGLADES **INSECTS**

Zebra longwing butterfly

Black salt march mosquito

Lubber grasshopper

invertebrate

An **invertebrate** is an animal without a backbone.

Birds

All of the animals of the Everglades have characteristics that give clues as to how scientists can **classify** them. The purple gallinule, or swamp hen, is covered in brightly colored feathers. Its large feet help it walk on floating plants.

The roseate spoonbill has pink feathers and two long legs. It looks for small fish, frogs, or other animals. When it finds one, it eats it with a snap of its spoon-shaped beak! Feathers, two legs, and beaks are characteristics that all birds have.

The purple gallinule also swims on the surface of water like a duck.

Why do you think this bird is called a spoonbill?

classify

To **classify** is to place into groups based on characteristics.

Reptiles

Alligators and crocodiles are the largest reptiles in the Everglades. Like birds, reptiles are **vertebrates**, or animals with **backbones**.

The Everglades is the only place in nature where alligators and crocodiles live side by side. How can scientists tell them apart? One way is to observe the snout, or front of the head. Both have nostrils on top of the snout so the animals can breathe while they float in the water. But observe how the shapes of the snouts are different.

TECHTREK
myNGconnect.com

Digital Library

The snouts of American crocodiles are more pointed.

American alligators have more rounded snouts.

vertebrate

A **vertebrate** is an animal with a backbone.

backbone

A **backbone** is a string of separate bones that fit together to protect the main nerve cord in some animals.

Fishes

The alligator gar is not an alligator at all, but a fish! Its thick scales, covers over its gills, and long snout must have made it look like an alligator to the people who started calling it that. The alligator gar is one of the biggest freshwater fish in North America. It can grow to be 3 meters (10 feet) and over 45 kilograms (100 pounds)!

The sawfish has more characteristics like a shark than a gar. Sawfish live in saltier water where the Everglades meets the ocean.

Alligator gars live further inland where the water has little salt in it.

Amphibians

The swamps and ponds of the Everglades make a great home for amphibians, such as the pine woods tree frog. Frogs, like almost all amphibians, live part of their life in water and part of their life on land. The pine woods tree frog tadpole lives in water. The adult frog spends time on land. This frog's sticky toes help it climb on palm fronds and tree limbs in this watery environment.

This young pine woods tree frog lives in water and has gills. It will grow lungs and legs later in its life.

The adult pine woods tree frog has legs and lungs like other amphibians. It also has sticky toe pads that help identify it as one of a group called tree frogs.

sticky toe pad

You can see that the American green tree frog also has sticky toe pads.

Mammals

Several mammals live in the Everglades. They all have hair or fur and make milk for their young that are born alive. But they can be very different from one another. Some eat plants, some eat insects, and still others eat other vertebrates. Only a few of some kinds are left, such as the Florida panther. It has characteristics much like panthers or cougars that live elsewhere. But the Florida panther lives only in southern Florida.

American mink

Southern flying squirrel

WHERE ARE THE FLORIDA PANTHERS?

Look at the brown area in the maps. A long time ago, large numbers of Florida panthers used to roam Florida and other parts of the southeastern United States. Today, scientists think only about 100 panthers are alive. They live in a small area of southern Florida, and their young are born only in the Everglades.

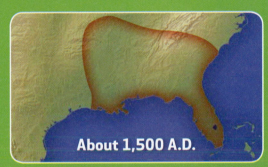

About 1,500 A.D.

Today

Young take about three months to develop inside the mother panther. She will have one to three kittens at a time. The kittens will drink their mother's milk for about a year.

CHAPTER 2 SHARE AND COMPARE

Turn and Talk How can the different kinds of animals in the Everglades be classified? Form a complete answer to this question together with a partner.

Read Select two pages in this section. Practice reading the pages so you can read them smoothly. Then read them aloud to a partner. Talk about why the pages are interesting.

my SCIENCE notebook **Write** Write a conclusion that tells the important ideas about what you have learned about animals in the Everglades. State what you think is the Big Idea of this section. Share what you wrote with a classmate. Compare your conclusions.

my SCIENCE notebook **Draw** Form groups of six. Have each person draw a different kind of animal that lives in the Everglades. Label the group you are drawing an example of. Label the characteristics you used to classify the animal. Combine the drawings to show the different kinds of animals in the Everglades.

CHAPTER
3

HOW DO PLANTS AND ANIMALS
LIVE IN THEIR ENVIRONMENT?

Look at the features of this environment. You see grasses and mountains in the distance. In summer, the weather is warm. The winter is cold and rainy. Deer do well in this place. This environment meets the needs of red deer.

TECHTREK
myNGconnect.com

Student eEdition | Vocabulary Games | Digital Library | Enrichment Activities

Red deer eat the grasses and shrubs that grow on the Scottish Highlands.

94   95

After reading Chapter 3, you will be able to:

- Recognize that environments support diversity of plants and animals.
  LAND ENVIRONMENTS, WATER ENVIRONMENTS

- Identify living things found in deserts, on grasslands, and in forests.
  LAND ENVIRONMENTS

- Identify living things found in salt water, fresh water, and wetland environments.
  WATER ENVIRONMENTS

- Identify and explain the flow of energy through a food chain.   **A FOOD CHAIN**

- Describe how a change in a food chain can be harmful and beneficial.
  ENVIRONMENTAL CHANGES

- Science in a Snap! Identify and explain the flow of energy through a food chain.   **A FOOD CHAIN**

HOW DO PLANTS AND ANIMALS LIVE IN

Look at the features of this environment. You see grasses and mountains in the distance. In summer, the weather is warm. The winter is cold and rainy. Deer do well in this place. This environment meets the needs of red deer.

THEIR ENVIRONMENT?

Red deer eat the grasses and shrubs that grow on the Scottish Highlands.

SCIENCE VOCABULARY

population
(pop-yūh-LĀ-shun)

A **population** is all the living things of the same kind that live in an environment. (p. 99)

A population of bison living together on the same prairie eat mostly grasses.

community
(kuh-MYŪ-nuh-tē)

A **community** is all the living things in an environment. (p. 100)

Many kinds of trees are part of the forest community.

food chain (fūd chān)

A **food chain** is a path by which energy is passed from one living thing to another. (p. 108)

The sun is the source of energy for a food chain.

my
Science Vocabulary

community (kuh-MYŪ-nuh-tē)

consumer (kuhn-SŪ-mur)

decomposer (dē-kuhm-PŌZ-ur)

food chain (fūd chān)

population (pop-yūh-LĀ-shun)

producer (pruh-DŪS-ur)

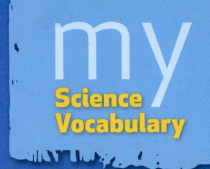

TECHTREK
myNGconnect.com

Vocabulary Games

producer (pruh-DŪS-ur)

A **producer** is a living thing that makes its own food. (p. 110)

A tree is a producer as it makes its own food.

consumer (kuhn-SŪ-mur)

A **consumer** is a living thing that eats plants or animals. (p. 111)

The bobcat is a consumer as it eats other animals.

decomposer (dē-kuhm-PŌZ-ur)

A **decomposer** is a living thing that breaks down dead things and the waste of living things. (p. 112)

Decomposers can break apart the wood and bark of a fallen log.

Land Environments

Deserts No area on Earth is as dry as a desert. Some deserts can be very hot in the daytime. Desert plants and animals can live with little water. For example, cactus plants store water in their thick stems. It is often hard to find food in a desert. A kangaroo rat stores food in its burrow to eat later.

 The Sonora Desert, in Arizona, is a good place for these saguaro cactus plants to grow.

The kangaroo rat gets the water it needs from its food.

Grasslands Grasslands are areas of flat land covered with mostly grasses. Few trees grow on grasslands. Animals, such as bison, live there. They graze or eat grass. All the bison in one area are called a ==population==. A population is a group of the same animals or plants that live in the same environment.

Other animals live on grasslands, too. Prairie dogs eat grass and live underground. Foxes and snakes hunt the prairie dogs.

TECHTREK
myNGconnect.com

Digital Library

These bison are living on a grassland in South Dakota.

Forests In areas where there is enough rainfall to grow trees, you may have a forest environment. Forests that grow near the Equator are usually rain forests. The plants in a rain forest grow well in a warm and wet environment. The other living things that live in a rain forest <mark>community</mark> grow well here, too. A community is all the living things in a certain environment.

 The Amazon rainforest is hot and wet. The world's largest rain forest has the most different kinds of plants and animals living in any place.

Scarlet macaws eat fruits and seeds that they find in trees.

Ocelots hunt at night. They eat lizards, frogs, and other small animals.

Forests also grow in areas with cooler temperatures. These environments usually have seasons. Some of these forests are made up of trees that drop their leaves in winter. Others are evergreen forests made up of pines and other conifer trees. These trees do not drop their leaves.

TECHTREK
myNGconnect.com

Digital Library

Deciduous forest

Coniferous forest

Before You Move On

1. What is a community? Describe an example.
2. Name two differences between a desert and a rain forest environment.
3. **Infer** Do you think a rain forest tree can grow in a desert? Why or why not?

Water Environments

Salt Water The ocean contains most of the water on Earth and it is salt water. The salt water environment provides the populations that live there with what they need. Most animals and plants live near the surface where sunlight reaches. You'll find fish, such as salmon and sharks, and mammals, such as whales and dolphins. Other animals, such as crabs and sea stars, also live there.

coral

fish

sea star

Where in the ocean can you find the greatest variety of living things? The answer is a coral reef. Coral reefs are made of the skeletons of tiny animals called corals. The coral feed on living things that need sunlight to grow. So coral reefs are only found in warm, shallow ocean water. The coral reef provides a place to live for many ocean animals.

Fresh Water Fresh water is water containing very little salt. Fresh water is usually found in rivers, streams, lakes, and ponds. It is also found in glaciers near the North and South poles. Fresh water in lakes and ponds is usually still or standing water. Standing water is a good environment for many types of plants and animals such as those in the pictures below.

Frogs often rest on the lily pads that float in ponds. Lily pads are the flat leaves of plants called water lilies.

Dragonflies lay their eggs in or near the water. They eat small insects that live in their fresh water community.

Beavers build their homes in fresh water. They eat wood from trees that grow nearby.

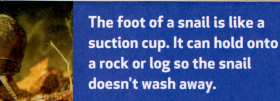
The foot of a snail is like a suction cup. It can hold onto a rock or log so the snail doesn't wash away.

Fresh water in rivers and streams flows or moves. Flowing water provides another type of water environment. Plants in a stream have strong roots that hold them in place. Some trees, such as willows, grow well near water. Trout, catfish, and other fish find food and shelter in the moving water. Snails, mussels, and crayfish live on the bottom.

Wetlands Wetlands, such as a marsh, can have either fresh water or salt water. A marsh is a wetland where tall grasses grow. A wetland is land soaked with water and often covered in water.

Some animals, such as muskrats, live near the water. You might see turtles, alligators, river otters, and shrimp. You'll also find large populations of mosquitoes. They live among the cattails, bulrushes, and grass.

This salt water marsh in Virginia forms where fresh water meets the salt water of the Atlantic Ocean.

Fiddler crabs live near the water's edge. They feed on marsh grasses and animals.

Egrets build nests among the marsh grasses. They wade in the water to catch fish.

A wetland where trees grow is called a swamp. Like a marsh, a swamp is sometimes covered in water. Many swamps form near slow-moving rivers. Most of the animals that live in marshes live in swamps also. You'll find alligators, crocodiles, bobcats, bears, beavers, insects, birds, and crayfish.

Cypress trees grow wide roots. They help hold the trees in the swamp.

Snapping turtles live in swamps.

Before You Move On

1. What is the differences between a salt water environment and a fresh water environment?
2. What is a wetland?
3. **Draw Conclusions** Why do ocean plants live close to the surface?

A Food Chain

What is in food that all living things need? The answer is energy. Plants get energy from the sun. Animals that eat plants get some of that energy. So do animals that eat other animals. A **food chain** is the path by which energy is passed from one living thing to another. Each living thing is a link in the chain.

Sunlight is the source of energy for most living things.

TECHTREK
myNGconnect.com
Enrichment Activities

A FOREST **FOOD CHAIN**

This diagram shows how energy from the sun can flow through a food chain.

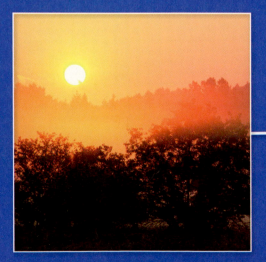

sun

oak trees

Every environment has food chains. A desert food chain might go from cactus to insect to lizard to hawk. An ocean food chain might go from seaweed to small fish to large fish to whales. A forest contains many living things, such as trees, bushes, grasses, insects, birds, and mammals. See how they can fit in a forest food chain.

gypsy moth caterpillar

robin

bobcat

Producers and Consumers Most plants get energy from the sun. They are called producers. **Producers** use the sun's energy to produce, or make, their own food. A plant uses some of the food it makes to grow. The rest of the food is stored in the plant's parts. In a forest, the largest producers are trees. Smaller producers include mosses, ferns, vines, and bushes.

PRODUCER Oak trees use energy from the sun to make food.

CONSUMER The gypsy moth caterpillar is an herbivore. It eats only plants.

CONSUMER The robin is an omnivore. It eats plants and animals.

CONSUMER The bobcat is a carnivore. It eats other animals.

Animals are **consumers**. They eat, or consume, other living things to get energy. Many animals, called herbivores, eat only plants. Other animals, called carnivores, eat only animals. Some animals eat plants and animals. They are called omnivores. For example a fox is an omnivore. It eats mice and other small animals. It also eats fruit.

Decomposers What happens when a plant or an animal dies? Dead things are not wasted. When they rot, their materials are used again. **Decomposers** break down dead things and the waste from living things. An earthworm is a decomposer. It helps return materials from dead things to the soil where plant roots take them in. The plant uses them to grow.

Earthworms are decomposers in soil. They break down tiny bits of dead plants and animals.

Fungi are decomposers. Many fungi break down dead wood or tree bark. You might see mushrooms on dead logs. These mushrooms are fungi. Beetles and other insects also break down dead things.

These mushrooms are decomposers. They are breaking down this dead log.

Before You Move On

1. What is the source of energy in most food chains?
2. How is a consumer different from a producer?
3. **Sequence** A rabbit, an owl, and some grass live in the same area. Write the order in which they would form a food chain.

Environment Changes

Changes in Food Chains Wolves and deer are part of a food chain in some forests. The deer eat plants, and wolves hunt the deer. In this food chain the deer are prey and the wolves are predators. The populations of wolves and deer affect one another. If the number of deer drops, then the wolves have less food to eat. Then the number of wolves drops, too.

This mule deer is on the alert for predators.

For many years people hunted wolves. Soon there were far fewer wolves. As the wolf population fell, the deer population grew. Why? There were not enough wolves to feed on the deer. People began to understand that wolves kept the number of deer from growing too large. In the forest environment, the deer and the wolves depend on each other. They keep their populations in balance.

Wolves are predators. They hunt and eat deer and other animals.

Links in a Chain

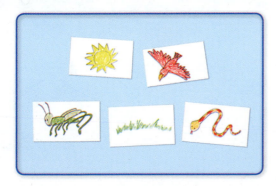

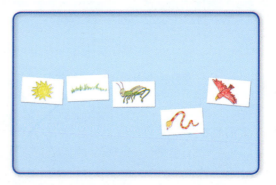

Draw links in a food chain from an environment you read about.

Put the cards in the correct order. Then remove a card.

Explain what will happen to the food chain and why.

Seasonal Changes In the fall, the Luangwa River in Zambia, Africa, nearly dries up. All of the hippos that live there cannot find plants to eat. They have to travel far from the river to find plants. Some nearly starve. Many won't survive the dry season.

In the dry season, the Luangwa River does not have much water.

There is plenty of water for plants to grow during the rainy season.

Hippos crowd together as the river dries up.

Then the rainy season begins. The river swells over its banks. The land floods. When the water drains from the land, grasses and plants start to grow again. The hippos and other animals now have plenty of water and enough new plants to eat.

Before You Move On

1. Why do wolves depend on deer?
2. How can rainfall affect animals in a food chain?
3. **Explain** How can a food chain change?

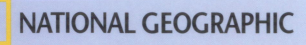
PROTECTING THE OCEAN ENVIRONMENT

People once fished the ocean with only boats and nets. Today, fishers find fish using technology. They use satellites, sonar, and aircraft. Are these tools working too well?

Overfishing in the Atlantic Ocean has reduced the numbers of many kinds of fish. Today the ocean has fewer tuna, swordfish, cod, and flounder. Some fish populations are now half as large as they were 50 years ago.

Wide nets help fishers catch large numbers of fish.

Some fishing practices damage the places where fish lay eggs and find food. Sometimes boats drag nets across the ocean floor catching many fish at once. However, the nets crush anything beneath them.

Too few fish is bad news for everyone. Some states have helped fish populations grow by protecting areas of the ocean. Sometimes fishing is not allowed. Other times fishing is allowed only at certain times. Either way, this allows the numbers of fish to increase.

Conclusion

Plants and animals live together in Earth's many environments. Plants are producers. Most plants make their own food using energy from the sun. Animals are consumers. They eat plants or other animals. Energy passes along a food chain from plants to animals. Decomposers break down dead things. They enrich the soil.

Big Idea Plants and animals depend on each other and their environment.

Producer

Consumer

Decomposer

Community

Vocabulary Review

Match the following terms with the correct definition.

A. community
B. consumer
C. decomposer
D. food chain
E. population
F. producer

1. All the living things in an environment
2. All the living things of the same kind that live in an environment
3. A living thing that makes its own food
4. A living thing that eats plants or animals
5. A path by which energy is passed from one living thing to another
6. A living thing that breaks down dead things and the waste of living things

Big Idea Review

1. Identify You see an earthworm, a pine tree, and a chipmunk. Identify each one as a producer, a consumer, or a decomposer.

2. List Write the names of two kinds of land environments and two kinds of water environments.

3. Compare and Contrast How are all consumers alike? How are they different?

4. Explain What happens to a food chain when one of its parts is destroyed?

5. Generalize Describe wetland environments and how they differ.

6. Draw Conclusions Grasses, zebras, and lions form a food chain. They live on the grasslands of Africa. What would happen if there were no rain and the grasses died?

Write About Food Chains

Explain How should these pictures be connected to show a food chain? Add a picture of your own to show a decomposer.

Sun

Chipmunk

Oak tree

Hawk

LIFE SCIENCE EXPERT: ZOO CURATOR

Maria Franke is the Curator of Mammals at the Toronto Zoo in Canada. She and her team raise young black-footed ferrets to return to the wild.

What do you do as part of the Black-Footed Ferret Recovery Implementation Team?

I oversee the breeding program for the black-footed ferrets at the Toronto Zoo. We raise young ferrets for release into the wild. I support the keepers who take care of the ferrets. I teach people about black-footed ferrets and raise money for the program. I was part of an international team that completed the first release of black-footed ferrets in Canada. We will watch the ferrets for several years to make sure they do well in the wild.

When you were younger, did you ever see yourself doing what you do now?

When I was young, I wanted to be a veterinarian. I became a zookeeper in Europe and Canada.

TECHTREK
myNGconnect.com

Student
eEdition

Digital
Library

What is a typical day like for you?

Every day is different. I make sure we have healthy populations of different animals. I help plan exhibits. I work with keepers to improve the care of animals. I give talks about the zoo's conservation work. I also work to save endangered animals. These include the Vancouver Island marmot, cheetah, and the rhinoceros.

What has been the coolest part of your job?

The coolest part is helping to save endangered species. Releasing black-footed ferrets in Canada was a real thrill. My job makes me feel like I am making a difference in the world.

TECHTREK
myNGconnect.com

Digital
Library

Releasing black-footed ferrets into the wild.

BECOME AN EXPERT

The Black-footed Ferret: Making a Comeback

In the 1800s, the grasslands of North America were home to a large **population** of black-footed ferrets. A population is all of the living things of the same kind that live together. But by the 1980s, almost all of the ferrets were gone.

The grasslands of North America cover much of the central United States and Canada. Black-footed ferrets and prairie dogs live on the grasslands.

population

A **population** is all the living things of the same kind that live in an environment.

Why did the population of black-footed ferrets almost disappear? The main reason was the prairie dog population got smaller. Prairie dogs are part of the grassland **community** .

community

A **community** is all the living things in an environment.

Ferrets and the Food Chain

Prairie dogs live in tunnels they have dug. The tunnels are known as burrows. Prairie dogs leave their burrows to eat grass. Black-footed ferrets also live in the burrows, where they hunt prairie dogs. A ferret may eat over 100 prairie dogs in a year. Grass, prairie dogs, and ferrets form a **food chain**.

Prairie dog burrows contain chambers for sleeping, nesting, and storing food.

food chain

A **food chain** is a path by which energy is passed from one living thing to another.

Almost all food chains begin with the sun. Grasses and other producers get their energy from the sun. Animals, such as prairie dogs, are consumers. They eat the grasses. Black-footed ferrets are also consumers. They eat the prairie dogs. Food chains also include decomposers. Decomposers, such as earthworms, break down dead things and waste from living things.

GRASSLAND FOOD CHAIN

| Sun | Grass | Prairie Dog | Black-footed Ferret |

producer

A **producer** is a living thing that makes its own food.

consumer

A **consumer** is a living thing that eats plants or animals.

decomposer

A **decomposer** is a living thing that breaks down dead things and the waste of living things.

Changes in the Food Chain

In the 1800s, farming and ranching began to change the grasslands. People plowed the land for farming. Burrows were buried. Many prairie dogs were killed. The prairie dog population got much smaller. Once the prairie dogs began to disappear, so did the black-footed ferret.

A combine cuts a wheat crop on land that used to be prairie.

Farming and ranching changed the grassland food chain. The grasses were removed so the prairie dogs had less to eat. With less food more prairie dogs died. Then the black-footed ferrets had less to eat so they died, too.

TECHTREK
myNGconnect.com

Digital Library

Young prairie dogs do not travel far from their burrow.

The Ferrets Return

Scientists once thought that the black-footed ferrets were all gone. Then in 1981, a farm dog found a black-footed ferret in Wyoming. This led to the discovery of a small group of ferrets. That small group, however, was almost entirely wiped out by disease. In 1987, scientists caught 18 ferrets and used them to start a ferret program. In this program, young ferrets are raised by scientists. Then they are returned to the wild.

U.S. POPULATION OF BLACK-FOOTED FERRETS

In 1984, disease killed all but 18 ferrets. People helped save the population.

500 —			**1000**
100 — **129**		**120**	
80 —			
60 —			
40 —			
20 —	**18**		
0 —			
1984	1987	1989	2008

Today ferrets have a bright future. About 200 young ferrets are freed into the wild each year. Many ferrets are placed in grassland reserves, where there are plenty of prairie dogs. These areas have no farms, ranches, or cities. The ferret population may never again be as large as it was in the 1800s. But now, the ferrets have places to live and grow.

Young black-footed ferrets are called kits. These kits were born at the Phoenix Zoo.

SHARE AND COMPARE

Turn and Talk How did the populations of black-footed ferrets and prairie dogs change over time? Form a complete answer to this question together with a partner.

Read Select two pages in this section. Practice reading the pages. Then read them aloud to a partner. Talk about why the pages are interesting.

my SCIENCE notebook **Write** Write a conclusion that tells the important ideas about what you have learned about how populations in an environment can change. State what you think is the Big Idea of this section. Share what you wrote with a classmate. Compare your conclusions.

my SCIENCE notebook **Draw** Work in teams to draw a picture of the grassland community in the 1800s or in the 1980s. Label the different kinds of living things there. Use arrows to show a food chain in the community. Compare your drawing with another team's drawing of the other time period. As a group, describe how they are alike and different.

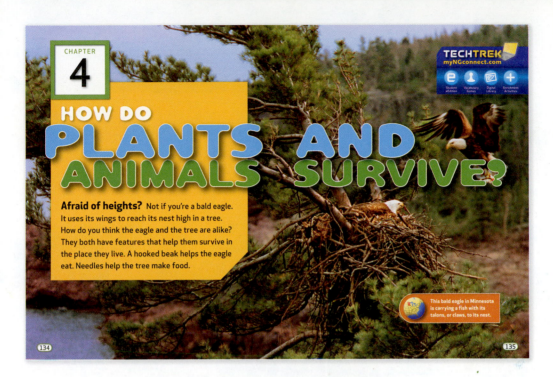

CHAPTER
4

HOW DO PLANTS AND ANIMALS SURVIVE?

Afraid of heights? Not if you're a bald eagle. It uses its wings to reach its nest high in a tree. How do you think the eagle and the tree are alike? They both have features that help them survive in the place they live. A hooked beak helps the eagle eat. Needles help the tree make food.

TECHTREK
myNGconnect.com

Student eEdition · Vocabulary Games · Digital Library · Enrichment Activities

This bald eagle in Minnesota is carrying a fish with its talons, or claws, to its nest.

134   135

After reading Chapter 4, you will be able to:

- Identify and describe the observable parts of plants that allow them to live in their environments. **PLANT ADAPTATIONS**

- Identify and describe the physical and other features of animals that allow them to live in their environments. **ANIMAL ADAPTATIONS**

- **Science in a Snap!** Identify and describe the physical and other features of animals that allow them to live in their environments. **ANIMAL ADAPTATIONS**

HOW DO PLANTS ANIMALS

Afraid of heights? Not if you're a bald eagle. It uses its wings to reach its nest high in a tree. How do you think the eagle and the tree are alike? They both have features that help them survive in the place they live. A hooked beak helps the eagle eat. Needles help the tree make food.

myNGconnect.com

Student
eEdition

Vocabulary
Games

Digital
Library

Enrichment
Activities

AND SURVIVE?

This bald eagle in Minnesota is carrying a fish with its talons, or claws, to its nest.

SCIENCE VOCABULARY

adaptation (a-dap-TĀ-shun)

An **adaptation** is a feature that helps a living thing survive in its surroundings. (p. 139)

The spines of a cactus are an adaptation for surviving in a hot, dry environment.

camouflage (CAM-uh-flahj)

Camouflage is an adaptation that allows a living thing to blend into its surroundings. (p. 148)

The snake's color and shape are its camouflage.

my Science Vocabulary

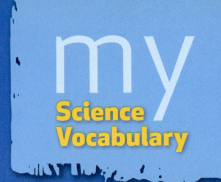

adaptation
(a-dap-TĀ-shun)

camouflage
(CAM-uh-flahj)

instinct
(IN-stinkt)

mimicry
(MI-mi-crē)

TECHTREK
myNGconnect.com

Vocabulary
Games

mimicry (MI-mi-crē)

Mimicry is an adaptation in which one kind of organism looks like another kind. (p. 148)

> This caterpillar uses mimicry as its tail looks like a snake's head.

instinct (IN-stinkt)

An **instinct** is an adaptation that an animal is born with and that controls its behavior. (p. 151)

> An instinct of the adult bird is to feed its young.

Plant Adaptations

Making Food and Storing Water Plants make food and store water. They do this in different ways depending on their environment, or where they live. Rain forest plants make food in their broad leaves. A broad shape helps the leaf take in sunlight. But it can lose a lot of water. Rain forest plants receive plenty of rain. They do not need to store water.

In the Amazon rain forest, the weather is hot and wet throughout the year.

This cactus grows in a hot and dry environment. It has **adaptations** for storing water. Its stem is thick and fleshy. The stem stores water and makes food. The leaves are also adapted for saving water. They are thin spines rather than broad and flat. They lose very little water.

The sharp spines of the cactus protect the plant from hungry animals.

spines

stem

Protecting If an animal bites a thorny plant, it will get a sharp stab from the thorns. Plants with thorns are often found in environments with many plant-eating animals. Thorns help these plants survive because the animals are less likely to eat them.

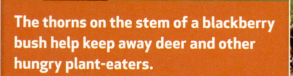

The thorns on the stem of a blackberry bush help keep away deer and other hungry plant-eaters.

Plants have other adaptations for protection, too. For example stone plants look like stones. The milkweed has a chemical that makes it poisonous to some animals. Tree bark is the outer layer of a tree's stem or trunk. The bark of some trees is so thick it can protect the tree from fire.

Why do you think the stone plant is hard for animals to see in a rocky environment?

Some sequoia trees have survived more than one forest fire because of their thick bark.

Reproducing Color and odor are adaptations. How do they help plants? They attract animals, and animals help in pollination. Pollination is the movement of a sticky powder called pollen from one flower to another. Most flowers need to be pollinated so the plant can reproduce.

Most of these colorful flowers, near the base of Mt. Rainier in Washington State, rely on insects to pollinate them.

Animals that spread pollen are called pollinators. Many pollinators visit flowers to drink nectar, a sweet liquid that flowers make. Like bright colors and strong odors, nectar is an adaptation of plants that attracts pollinators. Some adaptations attract a specific pollinator, such as an insect or bird, that lives in the plant's environment.

TECHTREK
myNGconnect.com

Digital Library

Bees are pollinators for many plants, including plants in the wild and farm crops. As the bee takes nectar from the chrysanthemum, what else might it be doing?

Before You Move On

1. What is an adaptation?
2. What adaptations help protect a plant?
3. **Compare** How are adaptations that protect a plant different from adaptations that help in pollination?

Animal Adaptations

Getting Food This young brown bear has just caught a fish, which it soon will eat. Just like plants, animals have adaptations that help them survive in their environment. Look closely at the brown bear. It has sharp teeth for catching and eating foods such as meat and fish. It also has flat teeth for chewing plants.

Brown bears use their teeth and claws to catch and eat fish and other animals.

claws

teeth

Animals have different adaptations for getting food, depending on what they eat. Look at this chameleon. Chameleons have long sticky tongues. How do you think the chameleon uses its long sticky tongue to help it survive? Like bears, chameleons also have teeth. Chameleons' teeth are small and are used to break up the insects they eat.

A chameleon can unroll its long tongue very quickly. This helps it surprise insects before they can get away.

Support and Movement

An owl is a hunter. It searches for and eats small animals, such as mice. Both hunters and the animals they hunt have adaptations that help them survive. An owl has large wings and a tail, and its bones are light. These adaptations make it a good flyer.

The four legs of a mouse help it run very fast. Because of its small size, the mouse can dart into a hole and escape.

TECHTREK
myNGconnect.com

Enrichment Activities

The owl's good eyesight helps it see the mouse against the white snow. The owl also uses other adaptations to find a meal.

Controlling Body Temperature Animals that live in cold environments have adaptations for keeping warm. Seals, for example, have thick layers of fat called blubber that act like a blanket against the cold.

Other animals live where the weather is hot. These animals often have adaptations for keeping cool. An elephant's big ears help it lose body heat on hot days.

Seals fold in their flippers to save heat. But when the weather warms, they flip them out to cool off!

This jackrabbit lives in a hot desert. What body part helps it keep cool?

Protection   How well can you see this snake? It is hard to see because of camouflage. Camouflage is an adaptation that allows a living thing to blend into its surroundings. The green snake is hard to see next to the green leaves.

Other animals have a form of protection called mimicry. Mimicry is an adaptation in which an animal actually looks like another plant or animal.

Find the bejuquillo snake in this photo.

The markings on the tail end of the spicebush swallowtail caterpillar look like a snake's head to other animals.

1. Draw an outline of an animal on colored paper. Cut out your picture. Place your picture on top of a piece of paper the same color as your animal. Stand back and observe your picture.

2. Now place your picture on a piece of paper that is a different color. Stand back and observe your picture.

Use your observations to explain how camouflage is a form of protection.

Raising Young Many animals do not care for their young. But other animals take care of their young for weeks, months, or years. Some animals build or find a home to raise their young. A fox moves into a den. The den provides shelter, or a safe place to live. Raising their young helps animals make sure that their offspring survive.

Fox kits depend on their mother for about 9 months. When the female kits are adults, by instinct they will make a den and feed their young.

The way some animals feed or raise their young is an **instinct**. An instinct is a behavior pattern that an animal is born with. They do not learn these behaviors. The behaviors are adaptations because they help the young survive.

This adult bird never learned that it should feed its young. It feeds its babies because of an instinct.

Before You Move On

1. What adaptations help grizzly bears get food?
2. What are two ways that animals protect themselves?
3. **Infer** Why do young fox kits need to stay in a den, but can leave it when they are older?

NATIONAL GEOGRAPHIC

LEARNING FROM ADAPTATIONS

Water is scarce in the outback of Australia, but the thorny lizard has an unusual way of getting a drink. The lizard can collect water on its skin. Channels run along the lizard's body. Water from a rain shower or even from wet desert sand collects in the channels and flows to the lizard's mouth.

Scientist Andrew Parker hopes to learn from the thorny lizard. He wants to build a machine to help people get water from desert sands, just as the lizard does.

The thorny lizard's ability to collect water might someday help people do the same thing.

Parker is hoping to follow other inventors who studied adaptations in nature. For example, plants such as burdock have a spiny, prickly seed pod. When a furry animal brushes against the seed pod, the prickles stick to the animal's fur. By studying the adaptations of burdock seed pods, a scientist invented Velcro®. This fastener mimics the prickles attaching to fur.

The two parts of Velcro® stick together to make a good fastener.

This close-up photo shows how the hooks and loops of Velcro® work.

Conclusion

Both plants and animals have adaptations, which are body parts or actions that help them survive. Adaptations may help an organism make or get food, protect itself from harm, or reproduce. Different plants and animals have a wide variety of adaptations to help them survive.

Big Idea Adaptations help plants and animals survive in their environments.

Animal adaptations

+

Plant adaptations

=

Help living things survive

Vocabulary Review

Match the following terms with the correct definition.

A. camouflage

B. mimicry

C. adaptation

D. instinct

1. A trait that helps a living thing survive

2. Something that an animal is born with that affects its actions

3. Something that allows a living thing to blend into its surroundings

4. Something in which one kind of living thing looks like another kind

Big Idea Review

1. List Name some adaptations plants use to protect themselves.

2. Recall What are some adaptations that help animals control body temperature?

3. Cause and Effect Sometimes the number of bees in an area increases right after a certain flower blooms. Why do you think this might happen?

4. Explain What kind of adaptations help a cactus survive in a desert?

5. Infer An insect called the walking stick looks just like a twig. What kind of adaptation is this? What does the adaptation tell you about the walking stick's environment?

6. Draw Conclusions Frogs have thin, damp skin. Could frogs survive easily in cold, snowy weather? Why or why not?

Write About Adaptations

Compare Observe these two insects. How are their adaptations alike? How are they different?

luna moth

praying mantis

CHAPTER 4

LIFE SCIENCE EXPERT: ZOOLOGIST

Kristofer Helgen

What do you do as a zoologist?

I oversee research in the world's largest museum collection of mammals, at the Smithsonian Institution in Washington, D.C. I work in museums around the world, studying mammals in order to learn how to tell thousands of mammal species apart. I also travel to tropical forests around the world to search for mammals in the wild.

When you were younger, did you ever see yourself doing what you do now?

Even when I was in elementary school, I knew I wanted to be a scientist who studies animals—a zoologist. I had never met any zoologists, but I had read about them in library books, so I knew that it was a "real job," something I could be when I grew up, if I worked hard enough.

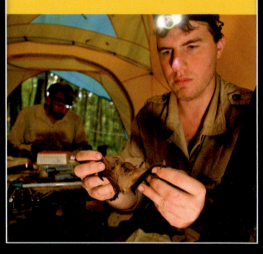

Helgen studies a blossom bat, which is an important pollinator in the rain forest of New Guinea.

Helgen observes a long beaked echidna in New Guinea.

TECHTREK
myNGconnect.com

e
Student
eEdition

Digital
Library

What did you study in school and college?

In school I was interested in almost every subject—I've always loved to read, and to learn! I was very interested in science and math, but also especially eager to learn as much as I could about geography. In college, I majored in biology, and this is where my serious training as a scientist really began.

What's a typical day like for you?

A typical day varies, depending where I am working. If I am in the natural history museum, I spend the day studying museum specimens, writing scientific papers, helping students and visiting scientists, and planning for museum exhibits. If I am on a field expedition, I spend the whole day (and night!) working to discover all the mammals that occur in the area where I am working.

Helgen looks for animals in a hollow tree.

BECOME AN EXPERT

Madagascar: Life on an Island

Millions of years ago, Madagascar was part of the rest of Africa. Then very slowly, it became an island and drifted to the east. Many plants and animals were trapped on the island. As the island's environment changed over millions of years, the plants and animals changed too. Today, Madagascar is home to plants and animals that live nowhere else on Earth. All have **adaptations** that help them live on the island.

Madagascar is an island about the size of Texas.

adaptation

An **adaptation** is a feature that helps a living thing survive in its surroundings.

Nymphalid butterfly

Sifaka lemurs live in Madagascar. They use their powerful hind legs to jump from tree to tree. They can leap over 9 meters (about 10 yards)! The sifakas' strong legs also help them hop quickly across the ground. Look at the mother and baby lemurs on this page. The baby knows how to cling to the mother's back because of an **instinct** . The baby does not have to be taught to hold on tight.

Sifaka lemurs have strong legs for leaping, and strong fingers and toes for gripping tree branches.

instinct

An **instinct** is an adaptation that an animal is born with and that controls its behavior.

Camouflage and Mimicry

Some animals of Madagascar are hard to find. Can you find the lizard in the photo? You need to look carefully! The lizard is hard to see against the bark of the tree. This is a form of <mark>camouflage</mark> , in which a living thing blends into its surroundings. Larger animals might eat the lizard if they could see and catch it. Because of its camouflage, the lizard is somewhat protected.

These scales of this lizard have the color and pattern of tree bark.

Camouflage

Camouflage is an adaptation that allows a living thing to blend into its surroundings.

Other Madagascar animals have a different adaptation for protection. Many frogs in Madagascar have bright colors. Some of these frogs are poisonous, but others are not. The ones that are not poisonous are mimics. By using an adaptation called **mimicry**, a harmless frog looks like another frog that may be poisonous. Animals that eat frogs learn to stay away from any colorful frog, whether or not it is poisonous.

The spots on the comet silk moth look like large eyes, so they scare away other animals.

The bright orange color of this harmless Golden Mantella frog mimics poisonous frogs.

Mimicry

Mimicry is an adaptation in which one kind of organism looks like another kind.

Baobabs and Fruit Bats

Several kinds of baobab trees grow in Madagascar. Like many trees, baobabs make flowers that depend on animals for pollination. Some of the Madagascar baobabs, such as Grandidier's baobabs, make large, white flowers. These flowers are pollinated by fruit bats, moths, and lemurs!

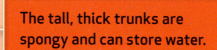

The tall, thick trunks are spongy and can store water.

The baobab's flowers have adaptations that attract fruit bats. The flowers open only at night, when the fruit bats often fly. Fruit bats visit the flowers to drink the nectar. As fruit bats fly from flower to flower, they carry pollen that gets stuck to their bodies. By spreading pollen, the fruit bats help the trees reproduce.

Like many bats, fruit bats hang upside down to rest.

The baobab tree has big white flowers that are easy to see at night.

CHAPTER 4

SHARE AND COMPARE

Turn and Talk   What kinds of adaptations do plants and animals have that live on Madagascar? Form a complete answer to this question together with a partner.

Read   Select two pages in this section. Practice reading the pages so you can read them smoothly. Then read them aloud to a partner. Talk about why the pages are interesting.

my SCIENCE notebook   **Write**   Write a conclusion that tells the important ideas about what you have learned about the adaptations of living things. State what you think is the Big Idea of this section. Share what you wrote with a classmate. Compare your conclusions.

my SCIENCE notebook   **Draw**   Draw a picture of a plant or animal on Madagascar and label its adaptations. Combine your drawing with those of your classmates to make a picture book of life on the island.

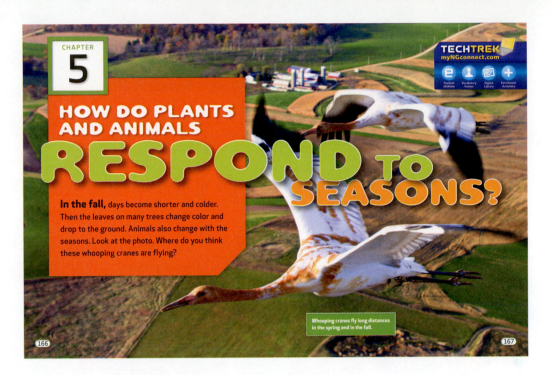

CHAPTER
5

HOW DO PLANTS AND ANIMALS RESPOND TO SEASONS?

In the fall, days become shorter and colder. Then the leaves on many trees change color and drop to the ground. Animals also change with the seasons. Look at the photo. Where do you think these whooping cranes are flying?

Whooping cranes fly long distances in the spring and in the fall.

TECHTREK
myNGconnect.com

Student eEdition | Vocabulary Games | Digital Library | Enrichment Activities

166

167

After reading Chapter 5, you will be able to:

- Compare how deciduous and evergreen plants change during the seasons.
  PLANTS CHANGE DURING THE SEASONS

- Explain how plants with different life cycles respond to changing seasons.
  PLANTS CHANGE DURING THE SEASONS

- Describe physical adaptations of animals that allow them to respond to changing seasons.
  ANIMALS CHANGE DURING THE SEASONS

- Describe behavioral adaptations that help animals survive changing seasons.
  ANIMALS CHANGE DURING THE SEASONS

- Identify adaptations that let plants and animals survive in places with wet and dry seasons. **OTHER SEASONAL CHANGES**

- **Science in a Snap!** Compare how deciduous and evergreen plants change during the seasons.
  PLANTS CHANGE DURING THE SEASONS

CHAPTER

5

HOW DO PLANTS AND ANIMALS RESPO

In the fall, days become shorter and colder. Then the leaves on many trees change color and drop to the ground. Animals also change with the seasons. Look at the photo. Where do you think these whooping cranes are flying?

TECHTREK
myNGconnect.com

Student
eEdition

Vocabulary
Games

Digital
Library

Enrichment
Activities

ND TO SEASONS?

Whooping cranes fly long distances in the spring and in the fall.

SCIENCE VOCABULARY

season (SĒ-zun)

A **season** is a time of year with certain weather patterns and day lengths. (p. 170)

> Winter is the coldest season.

deciduous (di-CIJ-yu-us)

A **deciduous** plant sheds its leaves every year. (p. 170)

> This deciduous tree has no leaves in the winter.

evergreen (EV-ur-grēn)

An **evergreen** is a plant that keeps its green leaves all year. (p. 171)

> These evergreen trees keep their thin leaves, or needles, through the winter.

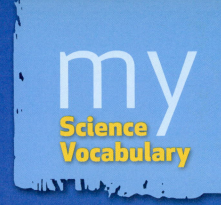

my Science Vocabulary

deciduous
(di-CIJ-yu-us)

evergreen
(EV-ur-grēn)

hibernate
(HĪ-bur-nāt)

migrate
(MĪ-grāt)

season
(SĒ-zun)

TECHTREK
myNGconnect.com

Vocabulary
Games

hibernate (HĪ-bur-nāt)

When animals **hibernate**, they go into a state that is like a deep sleep during cold winter months. (p. 177)

Chipmunks hibernate through the winter.

migrate (MĪ-grāt)

When animals **migrate**, they move to another place to meet their basic needs. (p. 178)

Monarch butterflies migrate in the fall.

Plants Change During the Seasons

Changes in the Way Plants Look Have you ever noticed how trees change during the seasons? A season is a time of year with certain weather patterns and day lengths.

Look at the winter scene below. The trees without leaves are deciduous. Deciduous trees shed their leaves in the fall. First, sugars and other nutrients in the leaves are moved to the rest of the tree. Then the leaves change color and fall to the ground. New leaves will grow in the spring.

Deciduous trees lose their leaves in fall.

Evergreen trees keep their green leaves all winter.

WINTER
Maple leaves would freeze in winter. Hard scales protect the buds.

SPRING
As the weather warms, new maple leaves grow from buds.

Evergreen trees do not lose their leaves every year. Many evergreen trees have long thin leaves called needles. A layer of hard wax protects the needles in winter.

Science in a Snap! Comparing Deciduous and Evergreen Leaves

aspen leaf

spruce leaves

Describe the shape of the aspen leaf and the spruce leaves. How are the aspen and spruce leaves similar? How are they different? Which of these leaves do you think is evergreen? Why?

SUMMER
Maple leaves stay green all summer long. They make food for the tree.

FALL
Maple leaves change color and dry out. Soon they fall from the tree.

Changes in the Way Plants Grow Many plants, such as cornflowers, live for only one year. In the spring, their seeds germinate, or begin to grow. During the summer, they flower and seeds begin to form. The plants die in the fall, but leave many seeds behind. Their seeds lie in the soil over winter. The seeds will begin growing in the spring.

cornflower

Some plants live for many years. Trees have woody stems that help them survive cold weather. Other plants have thin stems that die back in cold weather, but their roots stay alive. Lilies and tulips have bulbs. Bulbs grow underground. Bulbs store food that the plants use to begin growing in the spring.

A LILY THROUGH THE SEASONS

WINTER The plant survives cold weather as a bulb underground.

SPRING The bulb sends up a stem and leaves.

FALL Food moves down into the bulb and the leaves die.

SUMMER The plant flowers and forms seeds.

Before You Move On

1. What is an evergreen plant?
2. How do deciduous trees survive cold winter weather?
3. **Infer** A marigold dies after it makes seeds. Why might marigolds keep appearing in the same field?

Animals Change During the Seasons

Changes in the Way Animals Look To live through cold weather, many animals change in ways that help them keep warm. Deer, rabbits, and other animals grow heavy fur coats or thick layers of body fat. Some birds grow more feathers. When temperatures rise in the spring, the animals shed the extra fur or feathers. This helps them stay cool.

Musk oxen grow thick, heavy undercoats for winter.

In summer, musk oxen shed their undercoats.

Some animals change color with the seasons. In the summer, their fur or feathers are dark. This color blends in with the rocks, soil, and plants around them. It is hard for other animals to see them. In the winter, these animals turn white. Their white coats blend in with the snow.

TECHTREK
myNGconnect.com

Digital Library

In summer, arctic foxes have brown coats. In winter, they grow white coats. How does this change in color help the foxes live?

Changes in the Way Animals Behave It is difficult for many animals to find food in winter when leaves are gone and the ground is covered with snow. Some animals spend most of the fall gathering food and storing it. They eat that food during the winter.

Other animals eat extra food in the fall. This food is stored in their bodies as fat. Their bodies use the fat for energy during the cold winter. The extra layer of fat is also like a thick coat that keeps them warm.

This chipmunk is gathering food and will store it in its underground nest.

During cold winter months, some animals **hibernate**. When animals hibernate, they go into a state that is like a deep sleep. Their bodies use less energy when they hibernate. While they are asleep, the animals live on the fat stored in their bodies.

Many animals, such as chipmunks, find shelter and sleep much of the winter. From time to time they wake up to eat the food they stored in their nests.

TECHTREK
myNGconnect.com

Digital Library

In winter, chipmunks hibernate in their underground nests.

stored food

Some kinds of animals **migrate** when the seasons change. To migrate is to move to another place to meet basic needs. Many birds spend the summer in Canada and the northern United States. There they raise their young. In the fall the birds migrate to warmer places in the south. In the spring they return north. Migrating helps birds get the food they need.

Scarlet tanagers spend the winter in South America. In spring they return to the forests of eastern North America to raise their young.

Many insects live through the winter as eggs or larvae. Others burrow deep in the soil. But monarch butterflies migrate south in the fall. Moving to a warmer place helps the monarchs survive.

TECHTREK
myNGconnect.com

Enrichment Activities

Each fall monarch butterflies migrate south to California, Mexico, or Florida. In the spring monarchs fly north.

CANADA

UNITED STATES

MEXICO

Monarchs need to eat during their long flight south. This monarch is sipping nectar from a wildflower.

Before You Move On

1. List two ways animals may look different in summer and winter.
2. How does migrating help animals survive?
3. **Analyze** Some years oak trees do not make many acorns. Then there is less food for the animals. How might this affect an animal that hibernates?

Other Seasonal Changes

In many places, summers are warm or hot, and winters are cold. But weather does not change like this everywhere. For example, the African savanna has warm weather all year long. There the weather changes in other ways. Parts of the year are rainy, and other parts are dry.

In eastern Africa, great herds of wildebeests migrate in search of grass and water.

The baobab tree has leaves during the rainy season.

In the dry season, the baobab loses its leaves. This helps the tree save water.

During the rainy season on the African savanna, grasses and other plants grow well. Herds of wildebeests and zebras eat the grass. During the dry season, grasses do not grow. Then the herds migrate to other places where there is grass to eat and water to drink.

Before You Move On

1. How does the savanna change during the rainy season?
2. How are the seasonal changes of a baobab tree like those of a maple tree? How are they different?
3. **Predict** What would happen to the wildebeests if they did not migrate?

MANATEE MIGRATION

Manatees are sometimes called sea cows. Like cows, manatees are large animals that eat plants. They spend much of their time looking for plants to eat in shallow rivers, marshes, and ocean channels. Hydrilla and water hyacinths are some of the manatees' favorite foods.

Manatees can move easily between the fresh water of rivers and the salt water of the ocean. However, they cannot live in cold water. This is why many manatees migrate.

Even in Florida, the water temperature can rise and fall with the seasons. Manatees migrate in response.

Winter Range
■ Range

Summer Range
■ Range

Some Florida waters stay warm all year. Other Florida waters are too cool in winter for the manatees. Some manatees swim to the nearest power plant in search of warmer waters.

Power plants release warm water into the ocean. The manatees spend the cooler winter months in these warmer waters. At the end of the winter, the manatees migrate to their summer home.

Manatees are found in the ocean and rivers. They need warm water to live.

In the winter manatees gather in the warm water next to power plants.

Conclusion

Plants and animals respond to changing seasons. Deciduous plants shed their leaves during cold or dry seasons. Many animals grow and shed fur or feathers as the seasons change. Some animals change color. Some animals store food for the winter. Others eat extra food and build up fat. Some animals hibernate in winter. Others migrate as seasons change.

Big Idea Plants and animals have many different ways to live through changing seasons.

SOME PLANTS
- grow when the weather is warm
- shed leaves every year

SOME ANIMALS
- change how they look
- change how they behave

Vocabulary Review

Match the following terms with the correct definition.

A. evergreen

B. season

C. deciduous

D. hibernate

E. migrate

1. To go into a state that is like a deep sleep during cold winter months

2. A plant that keeps its green leaves all year

3. To move to another place to meet basic needs

4. A plant that sheds its leaves every year

5. A time of year with certain weather patterns and day lengths

Big Idea Review

1. Describe Describe two ways that an animal's body may change to help it stay warm in winter.

2. Identify What are the main seasons in the African savanna? In which season is there more food for wildebeests?

3. Explain How does changing color help an arctic fox live through the winter?

4. Sequence List these stages of a deciduous plant in order: leaves grow, branches are bare, leaves change color, buds open, leaves fall. Begin with winter.

5. Generalize Why is it more difficult for most animals to live in the winter than in the summer?

6. Predict People sometimes build roads or fences across animals' migration paths. How might these structures affect migrating animals?

Write About Seasons

Explain Daffodils grow and bloom in the spring. How does making a bulb help daffodils live through a cold winter?

Joe Duff

Operation Migration

Joe Duff helped found Operation Migration. Operation Migration uses ultralight aircraft to teach captive whooping cranes to migrate from Wisconsin to Florida.

What is your job with Operation Migration?

I am in charge of the team that trains the birds to follow our ultralight aircraft. As a pilot, I get to lead the birds on their first migration south.

What is the best part of your job?

Flying with the birds. They migrate in the fall when the colors are bright and the air is cool. We fly over hills and towns and down valleys. I can look out and see a long line of birds gliding off the wingtip. Then I realize how beautiful they are.

Have there been any surprises in your job?

No one could have predicted that we could use tiny airplanes to teach birds where to migrate. We were surprised when the first birds began to follow our aircraft. We were surprised when they came back on their own the next spring.

TECHTREK
myNGconnect.com

e
Student
eEdition

Digital
Library

What's been your greatest accomplishment?

In 2006, a whooping crane chick was hatched at the Necedah National Wildlife Refuge in Wisconsin. It was the first wild chick to hatch in the Midwest in over 100 years. That chick followed its parents to Florida along the route we showed them. Since then, it has returned every year.

A puppet is used to train a whooping crane chick.

TECHTREK
myNGconnect.com

Digital
Library

Whooping cranes follow an ultralight aircraft. The cranes are learning to migrate to Florida for the winter.

BECOME AN EXPERT

Shenandoah National Park:
Changing With the Seasons

Winter in Shenandoah Welcome to Shenandoah National Park! Here you will find tall mountains, rolling hills, and steep valleys. In Shenandoah, the weather changes from **season** to season. The changing weather affects all of the park's plants and animals. The photo shows the park in winter, when snow often blankets the land.

Shenandoah National Park is in the Blue Ridge Mountains of Virginia, not far from Washington, D.C.

season

A **season** is a time of year with certain weather patterns and day lengths.

Some animals, such as woodpeckers and squirrels, are active all winter. Bobcats hunt for rabbits, squirrels, and mice. The bobcats' thick fur keeps them warm.

In winter you would not see chipmunks or woodchucks. These animals , or go into a state that is like a deep sleep.

Bobcat

Pileated woodpecker

hibernate

When animals **hibernate,** they go into a state that is like a deep sleep during cold winter months.

Spring in Shenandoah

Spring brings warm weather to Shenandoah. Snow melts and rain falls. In the woods you can see wildflowers, such as violets, spring beauties, trilliums, and lady's slippers. You'll also see the new leaves of **deciduous** trees, such as oaks and hickories.

TECHTREK
myNGconnect.com

Digital Library

Many wildflowers, such as these yellow lady's slippers, bloom in the spring.

deciduous

A **deciduous** tree sheds its leaves every year.

Deer and many other animals of the Shenandoah have offspring, or young, during the spring. The young animals will find plenty of food to help them grow during the warm months ahead. Spring also brings the return of many birds. These birds **migrate** to warmer places during the winter. In the spring they come back to nest and raise their young.

The cerulean warbler spends the winter in South America. It returns to the Shenandoah in April or May.

new leaves

migrate

When animals **migrate,** they move to another place to meet their basic needs.

Summer in Shenandoah

Summer is the warmest season in the park. Plants grow quickly in the warm weather. The plants provide food to many animals. Deer nibble on leaves and stems. Bees, butterflies, and many other insects visit flowers, getting nectar and pollen.

Spotted skunk

Box turtle

Some animals, such as bats, owls, and skunks, come out at night. Skunks have a keen sense of smell that helps them hunt. They defend themselves with a foul-smelling spray.

Different plants and animals live in different parts of the park. In the Big Meadows region there are many ponds and marshes. During the summer, frogs hop in and out of the water, insects buzz overhead, and raccoons visit to catch fish.

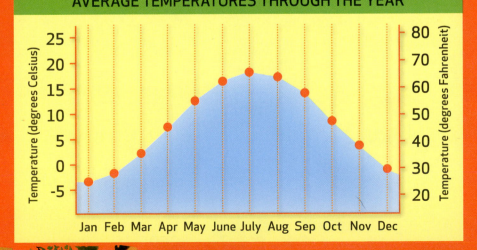

TEMPERATURES IN BIG MEADOWS

AVERAGE TEMPERATURES THROUGH THE YEAR

Temperature (degrees Celsius) — left axis: 25, 20, 15, 10, 5, 0, -5

Temperature (degrees Fahrenheit) — right axis: 80, 70, 60, 50, 40, 30, 20

Jan Feb Mar Apr May June July Aug Sep Oct Nov Dec

Fall in Shenandoah

When fall arrives, the weather becomes cool. The fruits of many plants are now ripe. Wild grapes hang from the vines. Many bushes have red or black berries.

The leaves of deciduous trees change color, and then fall to the ground. But the leaves of **evergreen** trees, such as pines, do not fall off. They will stay on the trees through the winter.

evergreen

An **evergreen** is a plant that keeps its green leaves all year.

The animals of the park are busy getting ready for winter. Squirrels gather hickory nuts and acorns. Bears eat fruits and acorns, as well as insects and fish. They store this food in their bodies as fat. Soon the bears will find dens where they will sleep through the winter. The fat in their bodies will keep the bears warm while they sleep.

When fall arrives, black bears store fat for winter. Then they find dens where they will sleep.

SHARE AND COMPARE

Turn and Talk How do the changing seasons in Shenandoah affect the animals that live there? Work with a partner to form a complete answer to this question.

Read Select two pages from this section. Practice reading the pages. Then read them aloud to a partner. Talk about why the pages are interesting.

my SCIENCE notebook **Write** Write a conclusion that tells the important ideas you learned about the seasons in Shenandoah. State what you think is the Big Idea of this section. Share what you wrote with a classmate. Compare your conclusions.

my SCIENCE notebook **Draw** Choose a season. Draw a plant or animal from the park as it may appear during this season. Add labels or write a caption to help explain what you drew. Share your drawing with your classmates. Group your drawings according to season.

Glossary

A

adaptation (a-dap-TĀ-shun)
An adaptation is a feature that helps a living thing survive in its surroundings. (p. 139)

B

backbone (BAK-bōn)
A backbone is a string of separate bones that fit together to protect the main nerve cord in some animals. (p. 51)

C

camouflage (CAM-uh-flahj)
Camouflage is an adaptation that allows a living thing to blend into its surroundings. (p. 148)

classify (CLA-si-fī)
To classify is to place into groups based on characteristics. (p. 50)

community (kuh-MYŪ-nuh-tē)

A community is all the living things in an environment. (p. 100)

consumer (kuhn-SŪ-mur)
A consumer is a living thing that eats plants or animals. (p. 111)

D

deciduous (dē-SIJ-yū-us)
A deciduous plant sheds its leaves every year. (p. 170)

decomposer (dē-kuhm-PŌZ-ur)
A decomposer is a living thing that breaks down dead things and the waste of living things. (p. 112)

E

environment (en-VĪ-ruhn-ment)
An environment is all of the living and nonliving things around an organism. (p. 20)

These pine trees live in a windy environment.

evergreen (EV-ur-grēn)
An evergreen is a plant that keeps its green leaves all year. (p. 171)

F

food chain (fūd chān)
A food chain is a path by which energy is passed from one living thing to another. (p. 108)

G

germinate (JUR-muh-nāt)
Seeds germinate when they begin to grow. (p. 21)

A bean seed can germinate if the soil is moist and warm.

H

hibernate (HĪ-bur-nāt)
When animals hibernate, they go into a state that is like a deep sleep during cold winter months. (p. 177)

I

instinct (IN-stinkt)
An instinct is an adaptation that an animal is born with and that controls its behavior. (p. 151)

invertebrate (in-VUR-tuh-brit)
An invertebrate is an animal without a backbone. (p. 53)

M

migrate (MĪ-grāt)
When animals migrate, they move to another place to meet their basic needs. (p. 178)

mimicry (MI-mi-crē)
Mimicry is an adaptation in which one kind of organism looks like another kind. (p. 148)

O

organism (OR-guh-niz-uhm)
An organism is a living thing. (p. 10)

P

pollen (POL-uhn)
Pollen is a powder made by a male cone or the male parts of a flower. (p. 24)

When bees visit flowers, pollen sticks to their bodies.

population (pop-yūh-LĀ-shun)
A population is all the living things of the same kind that live in an environment. (p. 99)

producer (pruh-DŪS-ur)
A producer is a living thing that makes its own food. (p. 110)

R

reproduce (rē-pruh-DŪS)
To reproduce is to make more of its own kind. (p. 24)

S

season (SĒ-zun)
A season is a time of year with certain weather patterns and day lengths. (p. 170)

spore (SPOR)
A spore is a tiny part of a fern or moss that can grow into a new plant. (p. 28)

V

vertebrate (VUR-tuh-brit)
A vertebrate is an animal with a backbone. (p. 52)

A backbone helps many vertebrates walk, run, fly, jump, or swim.

These redwood trees are producers.

Index

Credits

Front Matter

About the Cover (bg) Andrew Geiger/Riser/Getty Images. (t inset) Andrew Geiger/Riser/Getty Images. (b inset) Michael Nichols/National Geographic Image Collection. **ii–iii** Specialist Stock/Corbis. **iv–v** Ekspansio/iStockphoto. **vii** (t inset) Andy Rouse/Getty Images. **viii–ix** (bg) Dianne Christie/Getty Images. **vi–vii** (bg) David Aubrey/Taxi/Getty Images. **ix** (inset) Rick & Nora Bowers/Alamy Images. **x–1** Norbert Wu/Minden Pictures/National Geographic Image Collection. **2** (b inset) blickwinkel/Alamy Images. (t inset) Nigel Cattlin/Visuals Unlimited. **2–3** (bg) FLPA/Chris Mattison/age fotostock. **3** (b inset) Jim Brandenburg/Minden Pictures/National Geographic Image Collection. (c inset) Cornforth Images/Alamy Images. (t inset) Mark John Maclean/Shutterstock. **4** (bg) Mike Johnson Marine Natural History Photography. (inset) Brett Hobson.

Chapter 1

5 Purestock/Getty Images. **6–7** (bg) Purestock/Getty Images. **8** (b) Nigel Cattlin/Visuals Unlimited. (c) Jerry & Marcy Monkman/Animals Animals. (t) FLPA/Chris Mattison/age fotostock. **9** (b) Chris Lloyd/Alamy Images. (c) Vladimir Sazonov/Shutterstock. (inset) Visuals Unlimited. (t) Nigel Cattlin/Visuals Unlimited. **10–11** (bg) Ekspansio/iStockphoto. **12–13** (bg) Lijuan Guo/Shutterstock. **13** (b) Ottfried Schreiter/imagebroker/Alamy Images. (bc) Image Source Black/Alamy Images. (t) Veniamin Kraskov/Shutterstock. (tc) Dmitry Naumov/Shutterstock. **14** (b) Malcolm Romain/iStockphoto. (t) Malcolm Romain/iStockphoto. **15** Jim Richardson/National Geographic Image Collection. **16–17** (bg) Katarina and Marek Gahura/Shutterstock. **18** (t) Paolo Aguilar/epa/Corbis. **18–19** (bg) Kazuyoshi Nomachi/Corbis. **20–21** Nigel Cattlin/Visuals Unlimited. (r) Nigel Cattlin/Visuals Unlimited. **22** Runk/Schoenberger/Grant Heilman Photography/Alamy Images. **23** Nic Miller/Organics image library/Alamy Images. **24–25** (bg) Vladimir Sazonov/Shutterstock. **25** (bl) Nigel Cattlin/Visuals Unlimited. (br) Tramper/Shutterstock. (tl) Anest/Shutterstock. (tr) Visuals Unlimited. **26–27** (bg) Jerry & Marcy Monkman/Animals Animals. **27** (l inset) Larry Mellichamp/Visuals Unlimited. (r inset) Martin Shields/Photo Researchers, Inc. **28** (b inset) Andrew Syred/Photo Researchers, Inc. (t inset) Chris Lloyd/Alamy Images. **28–29** (bg) FLPA/Chris Mattison/age fotostock. **29** (inset) Wildlife GmbH/Alamy Images. **30–31** (bg) Francois Gohier/Photo Researchers, Inc. **31** (l inset) Layne Kennedy/Corbis. (r inset) Jonathan Blair/National Geographic Image Collection. **32** (inset) Phil Degginger/Alamy Images. **32–33** (bg) Jamie VanBuskirk/iStockphoto. **33** (inset) Jerome Wexler/Visuals Unlimited. **34** (inset) Courtesy of the Jane Goodall Institute. **34–35** (bg) Gerry Ellis/Minden Pictures. **35** (inset) Stefan Auth/imagebroker/Alamy Images. **36** (t) Scientifica/Visuals Unlimited. **36–37** (bg) Michael Nichols/National Geographic Image Collection. **37** (t) Scientifica/Visuals Unlimited. **38** (l) Kathy Merrifield/Photo Researchers, Inc. (r) Pfeiffer, J./Arco Images GmbH/Alamy Images. **39** Peter Griffith/Masterfile. **40** Christopher Talbot Frank/Ambient Images Inc/Alamy Images. **41** (b) Radius Images/Alamy Images. (t) Alex Skelly/Flickr/Getty Images. **42–43** Amadeo Bachar/National Geographic Image Collection. **44** Peter Griffith/Masterfile.

Chapter 2

45 David Aubrey/Taxi/Getty Images. **46–47** (bg) David Aubrey/Taxi/Getty Images. **48** (t) Tui De Roy/Minden Pictures/National Geographic Image Collection. **49** (b) Guillen Photography/UW/Bahamas/Alamy Images. (t) Mel Curtis /Photodisc/Getty Images. **50–51** (bg) Tui De Roy/Minden Pictures/National Geographic Image Collection. **52–53** (bg) Andy Rouse/Getty Images. **53** (inset) Guillen Photography/UW/Bahamas/Alamy Images. **54–55** (bg) Picturesbyme/Shutterstock. **55** (b) Frank & Joyce Burek/Photodisc/Getty Images. (bc) Digital Vision/Getty Images. (c) Sergey Toronto/Shutterstock. (t) Kevin Panizza/iStockphoto. (tc) WaterFrame/Alamy Images. **56** (l inset) Brian P. Kenney/Animals Animals. (r inset) Jake Holmes/iStockphoto. **56–57** (bg) blickwinkel/Alamy Images. **57** (inset) Amy White & Al Petteway/National Geographic Image Collection. **58** (inset) David Fleetham/Visuals Unlimited. **58–59** (bg) Mel Curtis /Photodisc/Getty Images. **59** (inset) Roberto Nistri/Alamy Images. **60** (b) Jack Goldfarb/Design Pics/age fotostock. (t) David Wrobel/Visuals Unlimited. **61** (bg) George Grall/National Geographic Image Collection. (inset) Digital Vision/Getty Images. **62–63** (bg) Creatas/Jupiterimages. **63** (b) Reinhard, H./Arco Images GmbH/Alamy Images. (t) McDonald Wildlife Photog/Animals Animals. **64–65** (bg) Juniors Bildarchiv/Alamy Images. **65** (inset) Greenberg, Phyllis/Animals Animals. **66** (inset) Jason Edwards/National Geographic Image Collection. **66–67** (bg) Bob Elsdale. **67** (bc) Creatas/Jupiterimages. (bl) Digital Vision/Getty Images. (br) Vinicius Tupinamba/Shutterstock. (tc) yuyangc/Shutterstock. (tl) Paddy Ryan/Animals Animals. (tr) PureStock/SuperStock. **68–69** (bg) Flame/Alamy Images. **69** (b) Flame/Alamy Images. (l) Alex Wild/Visuals Unlimited. (r) Art Wolfe/Photo Researchers, Inc. (t) Nature's Images/Photo Researchers, Inc. **70** (b) Suzanne L. & Joseph T. Collins/Photo Researchers, Inc. (l) Gregory K. Scott/Photo Researchers, Inc. (r) John Mitchell/Photo Researchers, Inc. (t) Harry Rogers/Photo Researchers, Inc. **70–71** (bg) Stephen Dalton/Photo Researchers, Inc. **72** (inset) Anup Shah/Nature Picture Library. **72–73** (bg) Jim Brandenburg/Minden Pictures. **74–75** (bg) Jason Edwards/National Geographic Image Collection. **75** (inset) The Photolibrary Wales/Alamy Images. **76** (inset) Ken Lucas/Visuals Unlimited. **76–77** (bg) Alexey Stiop/Alamy Images. **77** (inset) DK Limited/Corbis. **78–79** (bg) WaterFrame/Alamy Images. **79** (b inset) Roberto Sozzani © 2008. (c inset) Gary Cranitch/Queensland Museum. (t inset) Fred Bavendam/Minden Pictures. **80** (l inset) Stockbyte/Getty Images. (r inset) yuyangc/Shutterstock. **80–81** (bg) Gallo Images-Peter Lillie/Getty Images. **81** (inset) Gail Shumway/Getty Images. **82** (bg) Mehgan Murphy/Smithsonian National Zoological Park. (inset) Jessie Cohen/Smithsonian National Zoological Park. **83** (bg) Comstock/Getty Images. (inset) Jessie Cohen/Smithsonian National Zoological Park. **84–85** (bg) Jupiterimages. **85** (c inset) John Pontier/Animals Animals.

(l inset) Gordon Mills/Alamy Images. (r inset) Leighton Photography & Imaging/Shutterstock. **86** (bg) Medfrod Taylor/National Geographic Image Collection. (inset) Tom Vezo/ Minden Pictures/National Geographic Image Collection. **87** (l) Lynn M. Stone/Nature Picture Library/ Alamy Images. (r) jeff gynane/Alamy Images. **88** (b) SeaWorldCalifornia/ Corbis RF/Alamy Images. (t) Doug Perrine/Nature Picture Library. **89** (bg) Chris Mattison/ Alamy Images. (l inset) Barry Mansell/ npl/Minden Pictures. (r inset) Tim Laman/National Geographic Image Collection. **90** (b) Kim Taylor/Minden Pictures. (t) Don Johnston/All Canada Photos/age fotostock. **90–91** (bg) Mark Conlin/ Alamy Images. **92** Leighton Photography & Imaging/ Shutterstock.

Chapter 3

93 Mark John Maclean/Shutterstock. **94–95** (bg) Mark John Maclean/Shutterstock. **96** (b) Jochen Schlenker/ Photographer's Choice RF/Getty Images. (c) Galen Rowell/ Corbis. (t) Jim Brandenburg/Minden Pictures. **97** (b) Jon Faulknor/iStockphoto. (c) FLPA/age fotostock. (t) Antony Edwards/IC Images Ltd. **98** (inset) Anthony Mercieca/ SuperStock. **98–99** (bg) Natural Selection/Creatas Images/ Jupiterimages. **99** (inset) Jim Brandenburg/Minden Pictures. **100** (b inset) Pete Oxford/Minden Pictures. (t inset) DigitalStock/Corbis. **100–101** (bg) Galen Rowell/Corbis. (bg) FLPA/age fotostock. **101** (b inset) Comstock Images/ Jupiterimages. (t inset) Tim Fitzharris/Minden Pictures/ National Geographic Image Collection. **102–103** Specialist Stock/Corbis. **104** (b inset) Willem Kolvoort/Nature Picture Library. (cb inset) David P. Lewis/Shutterstock. (ct inset) Steven David Miller/Nature Picture Library. (t inset) David Kay/iStockphoto. **104–105** (bg) July Flower/Shutterstock. **106** (l inset) Gerry Bishop/Visuals Unlimited. (r inset) Lara Ellis/Alamy Images. **106–107** (bg) David Muench/Corbis. **107** (b inset) Erwin and Peggy Bauer/Bruce Coleman/ Alamy Images. (t inset) Joseph Sohm/Visions of America. **108** (l inset) Jochen Schlenker/Photographer's Choice RF/ Getty Images. (r inset) Antony Edwards - I C Images Ltd. **108–109** (bg) Jochen Schlenker/Photographer's Choice RF/Getty Images. **109** (c inset) Igorsky/Shutterstock. (l inset) Arco Images GmbH/Alamy Images. (r inset) Joe Austin Photography/Alamy Images. **110** (b inset) Joe Austin Photography/Alamy Images. (cb inset) Igorsky/Shutterstock. (ct inset) Arco Images GmbH/Alamy Images. (t inset) Antony Edwards - I C Images Ltd. **112** (inset) Vinicius Tupinamba/ Shutterstock. **112–113** (bg) Jon Faulknor/iStockphoto. **114–115** (bg) John E Marriott/All Canada Photos/ Photolibrary. **115** (inset) altrendo nature/Getty Images. **116** (b inset) Mark N. Boulton/Photo Researchers, Inc. (t inset) Christian Heinrich/imagebroker/Alamy Images. **116–117** (bg) Frans Lanting/Corbis. **118–119** Hein van den Heuvel/Cusp/Photolibrary. **120** (cl inset) Joe Austin Photography/Alamy Images. (cr inset) Vinicius Tupinamba/ Shutterstock. (l inset) Antony Edwards - I C Images Ltd. (r inset) Tim Fitzharris/Minden Pictures/National Geographic Image Collection. **120–121** (bg) Comstock Images/ Jupiterimages. **121** (cl inset) Gary W. Carter/Corbis. (cr inset) Bernd Kohlhas/zefa/Corbis. (l inset) Comstock Images/ Jupiterimages. (r inset) fotodolkens.nl/Alamy Images. **122** (b) D. Robert Franz/Bruce Coleman Inc. (t) Toronto Zoo. **123** Toronto Zoo. **124** (inset) Inga Spence/Visuals Unlimited. **124–125** (bg) Roberta Olenick/All Canada

Photos/age fotostock. **126–127** (bg) Konrad Wothe/ Minden Pictures/National Geographic Image Collection. **127** (cl inset) Inga Spence/Visuals Unlimited. (cr inset) Jim Brandenburg/Minden Pictures/National Geographic Image Collection. (l inset) Nadiya/Shutterstock. (r inset) USFWS/U.S. Fish & Wildlife, National Black-footed Ferret Conservation Center. **128** (inset) Creatas/Jupiterimages. **128–129** (bg) Fesus Robert/Shutterstock. **129** (inset) Raymond Gehman/National Geographic Image Collection. **130** (l inset) Joel Sartore/National Geographic Image Collection. (r inset) Jeff Vanuga/Corbis. **130–131** (bg) U.S. Fish & Wildlife Service. **132** Roberta Olenick/All Canada Photos/age fotostock.

Chapter 4

133 Jim Brandenburg/Minden Pictures. **134–135** Jim Brandenburg/Minden Pictures. **136** (b) Sexto Sol/ Jupiterimages. (t) Joseph Sohm; ChromoSohm Inc./Corbis. **137** (b) WizData, inc./Shutterstock. (t) Darlyne A. Murawski/ National Geographic Image Collection. **138–139** (bg) Warwick Lister-Kaye/iStockphoto. **139** (inset) Joseph Sohm; ChromoSohm Inc./Corbis. **140** (inset) Jacques Croizer/ iStockphoto. **140–141** (bg) Alchemy/Alamy Images. **141** (l inset) John Glover/Gap Photos/age fotostock. (r inset) DigitalStock/Corbis. **142–143** (bg) Cornforth Images/ Alamy Images. **143** (inset) Patrizia Tilly/Shutterstock. **144** (l inset) Annie Griffiths Belt/Corbis. (r inset) Purestock/ Getty Images. **144–145** (bg) Bernd Zoller/ imagebroker/ Alamy Images. **145** (inset) Mike Severns/Stone/Getty Images. **146–147** (bg) Joe McDonald/Corbis. **147** (b inset) Gerlach Nature Photography/Animals Animals. (t inset) PhotoDisc/Getty Images. **148–149** (bg) Sexto Sol/ Jupiterimages. **149** (t inset) Darlyne A. Murawski/National Geographic Image Collection. **150–151** (bg) Andrew Cooper/npl/Minden Pictures. **151** (inset) WizData, Inc./ Shutterstock. **152** Mitsuaki Iwago/Minden Pictures/ National Geographic Image Collection. **153** (bg) Robert Clark/National Geographic Image Collection. (l inset) Stocksnapper/Alamy Images. (r inset) Robert Clark/National Geographic Image Collection. **154** (l inset) Darlyne A. Murawski/National Geographic Image Collection. (r) Patrizia Tilly/Shutterstock. **154–155** (bg) Raymond Gehman/ National Geographic Image Collection. **155** (l) Kristjan Backman/iStockphoto. (r) Mark W. Moffett/National Geographic Image Collection. **156** (t, b) Tim Laman/ Tim Laman Photography. **156–157** (bg) Ulla Lohmann. **158–159** (bg) Suzanne Long/Alamy Images. **159** (b) Martin Harvey/Gallo Images/Corbis. (t) Thomas Marent/Minden Pictures/National Geographic Image Collection. **160** Thomas Marent/Ardea.com. **161** (b) Ken Lucas/Visuals Unlimited. (t) Arco Images GmbH/Alamy Images. **162–163** (bg) Dianne Christie/Getty Images. **163** (b inset) Daudet Andriafidison/ Madagasikara Voakajy. (t inset) Photograph by Karen Kennedy/ PhotoClassic & Safaripics. **164** Ken Lucas/Visuals Unlimited.

Chapter 5

165, 166–167 www.operationmigration.org. **168** (b) Eduardo Garcia/Jupiterimages/Getty Images. (c) Dennis MacDonald/ PhotoEdit. (t) Morten Hilmer/Shutterstock. **169** (b) Patricio Robles Gil/Minden Pictures/National Geographic Image Collection. **170** (l inset) Ned Therrien/Visuals Unlimited. (r inset) Monsoon Images/Photolibrary. **170–171** (bg) Eduardo Garcia/Jupiterimages/Getty Images. **171** (bl inset) Martin

B. Withers; Frank Lane Picture Agency/Corbis.(br inset) Marvin Dembinsky Photo Associate/Alamy Images. (tl inset) Ambient Ideas/Shutterstock. (tr inset) tomashko/Shutterstock. **172–173** (bg) Frank Krahmer/Corbis. **174** (inset) Alaska Stock Images/National Geographic Image Collection. **174–175** (bg) Morten Hilmer/Shutterstock. **175** (l inset) Norbert Rosing/National Geographic Image Collection. (r inset) Jim Brandenburg/Minden Pictures/National Geographic Image Collection. **176–177** (bg) Bruce MacQueen/Shutterstock. **178** (inset) Frank Lane Picture Agency/Corbis. **178–179** (bg) Patricio Robles Gil/Minden Pictures/National Geographic Image Collection. **179** (r inset) John A. Anderson/Shutterstock. **180** (bl inset) Beverly Joubert/National Geographic Image Collection. (br inset) DLILLC/Corbis Super RF/Alamy Images. **180–181** (bg) Art Wolfe/Stone/Getty Images. **183** (bg) Mark Conlin/Alamy Images. (l inset) Comstock Images/PictureQuest/Jupiterimages. (r inset) Michael Patrick O'Neill/Alamy Images. **184** (l) icpix_can/Alamy Images. (r) Jim Brandenburg/Minden Pictures/National Geographic Image Collection. **184–185** (bg) Michael S. Yamashita/National Geographic Image Collection. **185** (inset) Paul Tessier/iStockphoto. **186, 187** Operation Migration Inc. **188** (t) Dennis MacDonald/PhotoEdit.

188–189 (bg) Shenandoah National Park. **189** (l inset) William Leaman/Alamy Images. (r inset) Anna Henly/Workbook Stock/USGS Western Ecological Research Center. **190** (inset) Pat & Chuck Blackley/Alamy Images. **190–191** (bg) Rick & Nora Bowers/Alamy Images. **192** (b inset) Pat & Chuck Blackley/Alamy Images. (t inset) Tom Lazar/Animals Animals. **192–193** (bg) Pat & Chuck Blackley. **194–195** (bg) Pat & Chuck Blackley/Alamy Images. **195** (inset) Pat & Chuck Blackley. **196** Pat & Chuck Blackley/Alamy Images.

End Matter

EM1 Jerry & Marcy Monkman/Animals Animals. **EM2** (l) Nigel Cattlin/Visuals Unlimited. (r) Vladimir Sazonov/Shutterstock. **EM3** (bg) Peter Griffith/Masterfile. (inset) Mel Curtis / Photodisc/Getty Images. **EM6–EM7** Jupiterimages. **EM9** Thomas Marent/Minden Pictures/National Geographic Image Collection. **EM11** Bob Elsdale. **EM12** (l, r) Scientifica/Visuals Unlimited. **EM15** Creatas/Jupiterimages. **EM18** Buskirk/Alamy Images. **Back Cover** (bg) Andrew Geiger/Riser/Getty Images. (bl, br) Mike Johnson Marine Natural History Photography. (c) NASA's Earth Observatory. (tl) Brett Hobson. (tl) National Geographic Image Collection. (tr) Richard Herrmann/Oxford Scientific (OSF)/Photolibrary.